# EMPOWER
## TO
# ENSLAVE

*Decoding Intentions for Product Design and Demand Forecasting*

## DR RAHUL MIRCHANDANI

PARTRIDGE

**To order additional copies of this book, contact**
Partridge India
000 800 919 0634 (Call Free)
+91 000 80091 90634 (Outside India)
orders.india@partridgepublishing.com

www.partridgepublishing.com/india

*To Mom & Dad*

# CONTENTS

# ACKNOWLEDGEMENTS

The journey that I have undertaken to complete this Book is simply beyond words. I owe sincere appreciation to so many individuals and institutions, each of whom helped in their own unique ways as I crossed every milestone over the past decade.

I must begin by thanking my Doctoral Program Guide, Dr S R Ganesh. This journey began with you, sir. And I have had the privilege of having you with me ever since, every step of the way. I cannot imagine having reached this far without your encouragement, constant guidance and untiring effort. I needed the pushing, the smileys, your wit and your letters. I have learnt so much from you. You are simply amazing. Thank you so much.

Many thanks to Dr Bindi Mehta, Dr Vidya Naik, Dr H. Santhanam for your unstinting support throughout the Doctoral Programme. Thanks to all the Professors at the Narsee Monjee Institute of Management & Higher Studies for sharing their knowledge and moulding my research skills over the years.

I also wish to thank every individual who shared deep insights with me during the Research process. Thanks to all the respondents who spoke to me at length and without inhibitions. Thank you for your time, your hospitality, your patience and your frank opinions. Without your inputs, meaningfully completing this book would have been impossible.

Thank you to my Aries family, especially Shama, Jayapradeep, Omkar and Keertan, for pushing me to get this book done, offering to help proof read, edit, help develop the web app to use the algorithm proposed in this book, collect data, critique, or simply leave me alone in a quiet spot so I can write. Thank you to all my students and my friends for believing in me. Navita and Shailesh, your constant reminders over the years did not go in vain. Thank you for your energy and your attitude. You kept my spirits high, always. I needed that.

And above all, a big thank you to my family. I would not be here without each of you. Thanks for being the pillars of support that I needed throughout this journey. Thank you for understanding my odd hours, my crazy travel itineraries and my many moods. Thank you Armaan for knowing why Daddy needed to spend hours writing. Thanks Nitya for the patient waiting and being super-mom to Armaan whenever I was away. I would never have completed this without 'you' being 'you'. Thank you!

# PREFACE

Our world no longer resembles its own past. The dynamism in markets is arising not simply from the interaction of the individual structural components of the industry, but also from the industry "field" itself. The post-pandemic recovery and evolving consumer habits have set the 'industry's "ground" itself in motion' creating an environment that can be aptly classified as a *"turbulent field"*. Demand has become uncertain, product life cycles have shortened, and competition has intensified. In such a situation, understanding demand, planning demand and linking supply with demand is crucial.

*Habits create markets.* And with emerging megatrends indicating that consumers now think and behave very differently, this book begins with a trinity that links what, who and why people buy. It is changing habits that create gaps in the product mix that consumers now expect will be filled by newly minted products and services to satisfy their new needs.

*Pain creates demand.* It is this feeling of deprivation, of some other having a better deal, of observing a new technology simplifying lives or saving cost, that creates a new need for a product or service to be designed, specifically to address these pain points. Understanding consumer intent - why people buy what they buy – which includes a consumer's own understanding and prior observations, presents itself as a *proxy for past data*.

*Markets expect more.* Consumers expect products to be designed with their evolving habits and intentions at the core of the design philosophy. They crave curated offerings, a-la-carte pricing and value added features, leading to a feeling of empowerment. *'They made this with ME in mind'* is gratifying and leads to instilling pride in ownership and thus, consumer enslavement. We are all slavishly loyal to the brands we love and adore.

With a renewed understanding of the dynamics that drive markets and consumer intent, industry now needs to forecast demand for these new products and services, even in the *absence of past or reliable data*. Using complex econometric methods has shown limited success in such scenarios.

During my Doctoral Research, I made an attempt to map intentions and use this knowledge to construct a predictive model for forecasting demand. Extensive literature search did not uncover any such model and this book uses the outcomes of my research to set out a step-by-step methodology for managers to use to forecast demand in today's dynamic industry environments. The model will be especially useful to industries marketing specialty products or new product concepts where past demand data is unavailable, and the web of demand triggers is known but constantly changing.

The forecasting process in this book will identify and assess the individual impact of each factor affecting demand separately. Strategies may be formulated to 'influence the influencers' based on this information. Demand estimates would also assist in devising operational frameworks, planning inventories, increasing manufacturing and supply chain efficiency, etc. This book will therefore, add to the body of knowledge in the field of forecasting and aid planning within emerging markets.

I do hope that readers will also use the web application created for Managers to use the forecasting process and mind mapping algorithm, available at **EmpowerToEnslave.com**

# CHAPTER 1

# The Dynamic Market

A market is any platform, physical or virtual, where buyers, sellers and potential buyers interact with an intent to exchange goods, services, ideas, relationships, solutions, etc. for cash, kind or otherwise. Market dynamics are forever changing, with variables creating variety and chaos. Variables are diverse. Consumer intent is characterized from fickle impulse to deep thought.

**Markets are created by Habits**

Our habits define our behaviour. However, often times, we take our habits for granted. We rarely pause to consider how our habits create markets for products and services.

It is 6 am and I have a habit of going for a run. A whole host of products and services serve this daily habit. Beginning with my alarm clock, its plastic casing, metallic moving parts including all nuts and bolts, batteries, bed side shelf, wood, carpentry, active wear textiles, sports clothing, running shoes, cap or bandanna, wrist band, music player, USB disk and their components, charger, pedometer, sunglasses, tissues and napkins, stores to purchase all of these, transportation to the park, gardeners and park maintenance services, sprinklers, water lines, plant nutrition and plant protection products for the trees and plants, the list goes on. One habit thus has created a need and a direct market for more than twenty eight products and services. And this is just one of several habits and routines of a typical day.

Take another habit. Eating cereal for breakfast creates a market for corn kernels, its processing equipment, cardboard for packaging, ink for printing boxes, designers, nutrition experts, advertising professionals, billboards, delivery and logistics for the farm to home supply chain, milk, sugar, fresh fruit accompaniments, cereal bowls and ceramics, supermarkets and so on. Once again a single habit creates a market for many.

Ignoring habits is a recipe for a marketing disaster. A certain Indian soap brand, very popular among women, was attempting to tap into the male consumer market. It spent on creating a campaign for a new variant of the soap targeted towards men. It ran a campaign with the halo of celebrity endorsement of one of Bollywood's most popular (and expensive) film stars but featured this alpha male relaxing in a bubble bath filled with pink rose petals. The habits of a typical male consumer (which was the intended target audience for the message) was in stark contrast to this campaign imagery, creating an instant disconnect.

In comparison, an antiseptic soap brand attempting to leverage its brand beyond antiseptic liquids featured children and men playing in teams, being playfully aggressive and delightfully competitive. They return home messy and dirty, are met with the 'oh-no-not-again-but-why-am-I-not-surprised' eye roll look from their mother and wife at the door. They are despatched promptly to the shower with a bar of antiseptic soap, only to return refreshed, deodorised and hygienically cleansed. The everyday occurrences in the consumer's home directly echoed in this campaign. A chord was struck instantly because the consumer had very similar habits. The only difference was the suggestion of using the antiseptic bar of soap instead of their regular one, seamlessly injecting the new product concept into an existing routine.

Products designed to serve specific habits helps increase the speed at which consumers will accept and start using the products and services. The market for breakfast cereal in India has been a slow starter. The world's leading brands of cereal are yet to make a remarkable dent in India. This is not because there is anything wrong with their product. Their world class brands of 'healthy, nutritious, balanced, quick-to-use breakfast packed in a box' sell on Indian shelves as they do the world over. The typical consumer groups like Young urban professionals (YUPPIES), Double-Income-No-Kids couples (DINKS), children, time starved and running to work individuals exist in plenty in India, as in other markets where cereal brands have major markets. However, the missing link is the Indian breakfast habits characterised by a hot, heavy (sometimes spicy) cooked meal every morning. Cold cereal just doesn't make the cut – yet! Speed of adoption has hence been painfully slow and attaining critical mass is still many years away, because the product did not dovetail into a habit that existed. Rather it required consumers to change their habits.

Changing habits is not impossible to achieve. In the case of breakfast cereal, to encourage consumption, marketers distribute branded or white labelled versions of their products as part of their school outreach and community service programmes. Kids are served cereal at school and their marketers hope that by the time the students graduate, they would have become used to the habit of not having a typical Indian hot breakfast. Moreover, the cereal breakfasts are distributed in schools where the graduates and their families can later afford the changed habits. They are rarely distributed in low income and rural schools where the move from free (sometimes forced) consumption to paid purchase is doubtful.


**Demographics define Habits**

Since habits play a core function in creating markets, it becomes essential to understand the demographics that help define habits.

Demographics are a set of variables that assist in deconstructing consumer behaviour. I would describe myself as a forty year old male second generation entrepreneur, with a doctoral degree in management, living in one of the affluent suburbs of an Indian metropolis, having a teenaged son, married for almost 20 years, living in my ancestral family home, having a passion for world travel, teaching and outdoor adventure. This 'definition' of who I am helps explain why I am able to own two cars (living in a family owned home increases personal disposable income) and why I have accumulated over 1 million frequent flyer miles (extensive travel for work as an entrepreneur running a growing business and adventure/holiday travel as a family of three).

As clinical as it may seem, but changing any one variable will change my habits as a consumer. For instance, try substituting above place of residence from 'an affluent suburb of an Indian metropolis' to 'a rented apartment in midtown Manhattan, New York', keeping all other variables including my income as an entrepreneur constant. After converting my income into equivalent US Dollars and factoring in the cost of living in my own rented apartment Manhattan, New York versus living in a self-owned ancestral family home in a suburb of a city in India, my purchasing power will radically decline. Will I still remain a consumer for two luxury cars and frequent family holidays?

Once again, substitute 'teenaged son and married for 20 years' with 'three-year-old daughter and married for 5 years'. Consider how different my buying patterns would now be, despite all other demographic variables remaining unchanged.

Keeping a hawk eye on demographics will help decode a consumer's habits. It will help decide if consumers are in a position to understand, afford and appreciate the product or service on offer.

### Table 1. Demographic Variables help to Decode Habits

| Do Consumers Understand? | Can consumers afford? | Will consumers appreciate? |
|---|---|---|
| Highest Education level | Income | Place of work |
| Media exposure | Savings | Place of residence |
| Number of years of schooling | Personal disposable income | Business Networks |
| Languages understood | Purchasing power | Personal circles |
| Reading habits | Number of dependent kids | Age |
| Family life cycle | Number of dependent parents | Gender |
| Other substitutes for the product currently being used | Number of working members in the family | Commuting time and Distance to/from work |

## Products Designed for Habits

Product design requires access to (or the creation of) technology, robust processes and appropriate manufacturing methods. A choice will need to be made on which features to bundle into the product being designed. Design choices will also decide the cost of the product's bill of materials and also the product's presentation to consumers.

A product can be as simple or as complex as its creator intended. Technology will decide the interface with which the consumer will touch, feel and use the product. However, all demographic groups will not be able to afford every piece of technology on offer. Some may find it complicated. Some may believe certain features are unnecessary and discard the product's design as excessive. Thus, demographic variables will help segment what consumers expect to see in the product being offered to them.

Look at the erstwhile Nokia 1100 phone models. The product was designed for a specific segment that included India's multitude of truck drivers. The phone was stripped down to basics – it would make a call, had a rugged exterior that wouldn't break with every fall, dust proof cover to suit the rural environment, long battery life and a torchlight that was usable during common power outages and for use while stopped on a dark highway. It did not have a fancy high megapixel camera, did not have smartphone features, could not access the internet, did not have a touch screen, and no such features which its user – the truck driver – would neither use nor appreciate. The product was not low on quality but was redesigned to include restricted features thus bringing down its market price significantly. Less was being bundled in to align with the target consumer's habits and hence less was being charged.

If a truck driver was told that he had to now pay double the price, because the phone came with a 18-megapixel front, selfie camera with a forward flash, this feature would be considered an unnecessary 'cost adding feature' as it would have very limited usage among truck drivers driving for days on dusty roads. They rarely stop on the highways for selfies.

Microsoft Office is a computer program that many use on a daily basis. It is feature abundant and even a regular user utilizes only a fraction of the available options and features. Many are unaware that some features even exist and most others see no utility for them. However, the price of the software includes the cost of every feature bundled in, whether a consumer uses it or not. Imagine a version of Microsoft Office with only the bare basic features that the average user utilizes. Its price would drop dramatically and consequently, would widen its user base. 'Student Versions' of the software attempt to do exactly this. And 'value for money' is created by restricting the feature bundle.

**Cost adding versus Value adding Features**

Cost adding features are features that a consumer rarely (if ever) uses or features which the user does not understand nor need. Its presence in the product's design adds to cost but the user does not perceive an equivalent value. 'Feature rich' thus becomes a source of avoidable waste. In contrast, value adding features are ones that a customer routinely uses or enjoys having in the product. He is therefore willing to pay for them, sometimes even at a premium.

While travelling on a two hour domestic flight, our intent is to get from origin to destination. To break down the airline travel 'product' into distinctive features helps deconstruct this concept.

**Table 2. Deconstructing the product**

| Value adding features in an airline product | Cost adding features in an airline product |
|---|---|
| On time performance | Four course meal service |
| Safety | Checked in baggage in the aircraft luggage hold |
| Security | Lounge access at the airport |
| Comfortable seats | Business class or Premium cabin |
| Efficient ground handling | |
| Space for carry-on baggage inside the cabin | |

For the airline passengers, what most airlines the world over have therefore done is to take away all the cost adding features from their product and retain only the value adding features. However, since some travellers are in the habit of travelling with heavy bags or would like to have a meal on board, these are offered to them using additional 'pay for use' pricing. Airlines therefore have taken away cost adding features, unbundled them and offered them back at a price, creating a second revenue stream, while reducing the basic 'seat-on-a-plane-that-gets-you-there' price.

## Reduced Cost by Reducing Cost Adding Features

Various companies across different sectors have successfully implemented this strategy to offer more affordable products while still maintaining a competitive edge in the market.

**Automobiles:** One of the most popular examples is the Toyota Corolla. Over the years, the Corolla has undergone numerous redesigns, and one of the key strategies employed by the company to reduce costs has been simplifying the car's features. By removing unnecessary options and features, Toyota has managed to make the Corolla more affordable, without compromising on its reliability and performance.

**Furniture:** IKEA, a well-known furniture retailer, has built its brand around offering affordable, stylish, and functional furniture. One of the ways they achieve this is by reducing the number of features and offering Do-It-Yourself assembly for their products. This allows them to cut production costs and pass the savings on to the consumer.

**Personal Care:** The Dollar Shave Club's business model is based on offering a subscription-based service for razors and grooming products. By removing the need for consumers to purchase expensive, brand-name razors, they have created a more affordable option for customers. The company achieves this by providing a limited selection of razors and blades, focusing on quality and simplicity.

**Apparel:** Uniqlo, a Japanese clothing brand, is known for its simple, minimalist designs. They have managed to reduce costs by focusing on a limited range of high-quality, basic clothing items, such as T-shirts, sweaters, and jeans. By eliminating unnecessary features and keeping their designs streamlined, Uniqlo offers affordable, stylish clothing options to consumers.

## Stick to the Core

The lack of certain habits therefore allows for customization of a feature rich product into a frugal, bare basics version that would discard cost adding features and retain only value added features. Thus, the variant would align with the target group's

**Figure 1. The Trinity that drives Market dynamics**

habits and they would believe – and rightly so – that the product was re-designed for them, creating a lasting feeling of empowerment. 'They made this product keeping me in mind' is a key to gaining long term consumer loyalty. Paying customers love being at the centre of the seller's universe. Who doesn't like to feel that everything revolves around them!

**In order for a market to thrive, <u>all three</u> forces of Habits, Technology and Demographics <u>must</u> fall in place <u>simultaneously</u>. Else, the market <u>will die</u>.**

Earlier, Breakfast Cereal and Cellular phones were introduced as examples. Had the cellular phone instruments been highly priced and had call rates not dropped from the initial Rs 16 per minute to the current levels of a few paise per minute, the technology would have remained out of reach due to unaffordability. The product would never have created the market depth- like it has today in India, with more Indians forecast to have access to cellular devices than even safe, drinking water. Though the habits of wanting to communicate and the technology to create a cellular phone both available, only a very limited demographic set would have used the technology, had high prices remained the norm. Since prices fell and became affordable to many, demographic variables aligned and the mobile telephony market now thrives.

In case of breakfast cereal though, the habits of cold cereal as breakfast is far from commonplace in India. Hence this market remains shallow. And the cultural preference for a hot cooked morning meal will keep the behavioural inertia (a.k.a. resistance to change) from breaking for a long time to come. Alternative segments are being specifically curated by Kellogg. They have made attempts to position the product as assisting weight loss and have also shown it as being a perfect combination with fresh fruit and milk. Most recently, the cereal has even been packaged for use in Mumbai's radio taxis, where executives commuting to work, stuck in a cab for a good amount of time each day get an option to while away their time in traffic munching on healthy cereal. If you are sure that you have a great product, identify the right demographic target and start working, like Kellogg, on changing their habits. The third piece of the puzzle will eventually click into place.

**Responding to Rapidly Evolving Consumer Habits**

Consumer habits are continuously evolving. This has led to a significant impact on the products available in the market. Companies have to adapt and innovate to meet the changing demands of their customers. Here are a few examples of how changes in consumer habits have forced changes in products.

**The Rise of Environmental Consciousness**

In recent years, consumers have become more aware of the environmental impact of their choices. As a result, there has been a shift towards sustainable and eco-friendly products. For instance, companies have started producing reusable shopping bags, biodegradable packaging, and energy-efficient appliances. This change in consumer habits has forced companies to reevaluate their product offerings and focus on creating environmentally responsible products.

## The Growth of Health-Conscious Consumers

The increasing awareness of the importance of health and wellness has led to a change in consumer habits. People are now more focused on consuming healthier food options, opting for organic and non-GMO products. This shift has forced companies to adapt their product offerings to cater to this new demand. For example, food manufacturers have started producing healthier snacks, low-sugar beverages, and organic food products.

## The Emergence of Digital Technology

The rapid advancement of digital technology has significantly influenced consumer habits. With the rise of online shopping and digital communication, consumers now expect products and services to be easily accessible and user-friendly. This has forced companies to adapt and improve their products to meet these expectations. For example, many businesses have shifted to online platforms, offering digital products and services, and implementing user-friendly interfaces for their websites and applications.

# CHAPTER 2

# Why Do People Buy What They Buy

It was New Year's Eve 1880. A globe of hollow glass with a thread like filament in the centre was the focus of everyone's attention. The gathering suddenly fell silent when the host stepped on to the make-shift dais at the centre of the room. Heated anticipation almost warmed the chill winter air. The silence was broken with the click of a switch. The filament within the hollow globe instantly turned incandescent. The electric current had heated the thread to emit light. The world would never be the same again. Edison had just made the first public demonstration of the light bulb. The news travelled fast. A new invention promised the end to the age of illumination by wax candles and the dependence of mankind on sunlight. Almost overnight, aspirations changed. Work no longer needed to stop at dusk. Common people who until then had accustomed themselves to the flickering flame of the candlestick suddenly expected more. The demonstration had unleashed an inner feeling of deprivation. It had created 'pain' and an "I can get something better" attitude that led to sale of the first electric bulb and every single one in the decades that followed until today. *It is this inner 'pain', an emotional feeling of dissatisfaction, which is every marketer's dream.*

**The Pain of Unfulfilled Desires**

Every one of us succumbs to this emotion fuelled by our "unfulfilled desires". To satisfy the desire, we search for possible 'satisfiers' to address our wants and consequently, we buy. When the want is satisfied, we desire more. This is perhaps laying the foundation for creating demand for any product or service - the selfish, human urge to have it all.

Flip the pages of any newspaper or magazine. Glance at hoardings that scream at you while you drive. Marketers almost fall over each other trying to make you feel deprived. They almost torment us with their tall promises and brainwash us with their messages. Consider some striking examples.

A certain hotel chain runs a campaign that says "Living in a Dream". Who does not want to live in a dream? The emotional connect is instant. The message goes on to say "Directed by Impulse, Starring You". Desire is unleashed. 'Pain' intensifies seeing the visual portrait of a picture perfect stretch of beach. We look around our immediate surroundings and ask ourselves, "What am I doing here? I would rather be there, now!"

A leading international airline's global television campaign asks viewers "When was the last time you did something for the first time?" Visuals show corporate executives taking a few moments to get drenched in the rain, an elderly lady traveling for the first time in a helicopter and ends with a little girl expectantly looking out of the picture windows at an airport watching planes taxi past. The almost child-like curiosity within all of us surfaces. Action is prompted when the airline logo appears with the simple punch line "Keep discovering." Almost instantly, our mind says "Let's get out and discover the world."

## Consumers' 'I Wish…' statements provide market opportunities

A look at some common 'Pain Points' related to certain product categories will indicate how these give rise to gaps and areas of improvement for newer brands in the category to enter or existing ones to upgrade.

### I wish my Smartphone had better Battery Life

One of the most common pain points associated with smartphones is battery life. Users often struggle with their devices running out of power too quickly, especially when using power-intensive features like GPS, high-resolution screens, and applications. This issue has led to the development of products such as portable power banks and fast chargers, which extend the battery life of smartphones and help users stay connected for longer periods.

### I wish my laptop had better Heat Management

Laptops, while convenient and portable, can generate significant heat due to their compact design and limited ventilation. This heat can cause discomfort for users and may potentially damage internal components. To address this pain point, companies have introduced products such as cooling pads and thermal compounds that help dissipate heat and maintain optimal temperatures for improved performance and user experience.

**I wish my Headphones had better Audio Quality**

Many users experience dissatisfaction with the audio quality produced by their headphones, often due to poor sound isolation, bass response, and overall clarity. To combat these issues, various products have emerged, such as noise-cancelling headphones, in-ear monitors, and high-quality speakers that deliver superior audio performance and an enhanced listening experience.

**I wish my Gaming Console had wider Compatibility**

Gaming consoles often face compatibility issues with older games or specific peripherals, which can limit the gaming experience for users. To address this pain point, companies have introduced products like retro gaming consoles, emulators, and adapters that enable users to enjoy their favourite games on modern platforms, without compromising on the nostalgic experience.

**I wish my Smart Home Device addressed all my privacy concerns**

While smart home devices offer convenience and efficiency, many users have privacy concerns regarding their data and security. To alleviate these concerns, products like smart home privacy screens and encryption solutions have been introduced, ensuring that users can enjoy the benefits of smart home technology without compromising their privacy.

**I wish my VR Headset was more comfortable**

Virtual reality (VR) headsets can cause discomfort for users due to their design and weight, which can lead to issues such as headaches and nausea. To address this pain point, products like adjustable head straps, cooling masks, and ergonomic designs have been introduced, providing a more comfortable VR experience for users.

## Messaging to Frustrate

Marketers have been able to sell us products whose need was perhaps never perceived until even a few years ago. Life was perfect even before the age of the cellular phone. People were in touch with their friends, family and business associates. It was business as usual, until the day when we were presented with this wonder of technology that provided us with the convenience of staying in touch even while on the move. All of a sudden, the mobile phone became a prized possession. Anyone who did not have one felt deprived, detached and disconnected.

Aspirations rose and mobile telephone companies focused all their energies on telling us how indispensable their service was. Overnight, every call we missed while we were travelling frustrated us even more. Suddenly, we wanted the convenience of making a call even on a remote vacation island. Pressure mounted seeing colleagues chatting away on their cell phones. Desire peaked and demand rose. Over time, habits changed. A cell phone became a necessity.

Today, every purchase we make is prompted by such 'inner pain.' Think about it. The first piece of clothing was purchased because it provided protection and addressed demands of propriety within the community. In certain cultures, it is known to have been used to indicate social status.

Even something as basic as a match box was successful as its use allowed human beings to dispense with the inconvenient, time consuming process of making fire by friction. Once a group of consumers made fire with the easy flick of a match, it hurt non-users to waste time and energy endlessly rubbing flint to serve the same purpose. It is to get rid of the feeling of deprivation that we do things like building homes and buying cars, use banks and purchase insurance policies.

Some advertisements are designed to make consumers feel frustrated because of not owning a product.

**Apple iPhone Ads:** Apple has been known for their effective marketing strategies, and their iPhone ads often create a sense of frustration among consumers who do not own their latest models. These ads showcase the newest features and innovations, creating a desire for the product and making those who do not have it feel left out. For example, the "Shot on iPhone" campaign highlights the impressive camera capabilities of the latest iPhone models, making users of older iPhones or other smartphone brands feel frustrated that they are missing out on capturing high-quality photos and videos.

**Luxury Car Ads:** Luxury car advertisements often target consumers' desire for status and exclusivity, leading to feelings of frustration among those who do not own such vehicles. These ads portray a luxurious lifestyle, featuring sleek designs, advanced technology, and superior performance. The emphasis on these desirable aspects can make viewers who do not own luxury cars feel frustrated or dissatisfied with their current vehicles. For instance, BMW's "Ultimate Driving Machine" campaign focuses on the brand's exceptional engineering and driving experience, creating a sense of longing among those who do not own a BMW.

**Fashion and Beauty Advertisements:** Fashion and beauty advertisements frequently employ tactics that make consumers feel frustrated if they do not possess certain products or follow specific trends. These ads often depict idealized images of beauty or fashion-forward individuals, enticing viewers to strive for that same level of attractiveness or style. For example, cosmetic ads may showcase flawless skin achieved through the use of their products, leaving those without them feeling frustrated about their own imperfections. Similarly, fashion ads may feature trendy clothing items that create a fear of being left behind in terms of style.

In all these examples, advertisers intentionally create a gap between those who own the product and those who do not, generating frustration among the latter group. By highlighting the benefits, features, and desirability of their products, advertisers aim to tap into consumers' emotions and create a sense of longing or dissatisfaction.

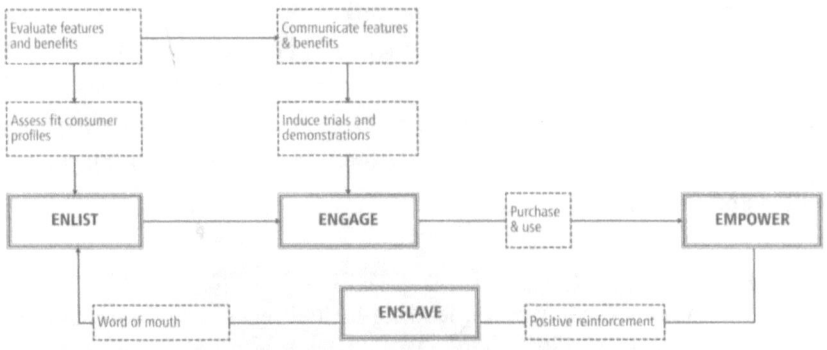

**Figure 2. The E's of the Marketing Process**

Marketers systematically' exploit our inner desires. Identifying the features and benefits provided by the product or service on offer is a crucial step in the overall marketing process. It lays the foundation for prospecting. Profiles or segments of potential buyers are drawn up on the basis of the needs that can be addressed by the offering. Prospective customers are **enlisted** and positioning strategies are developed to suit this target audience.

The next step is to **engage** the prospective customers. This is achieved by seamless, clear communication with the potential users. Messages should be direct and focused. Visuals are perhaps the most powerful way to engage a potential user and generate interest in the product on offer. Show them what they are missing. Marketers also try to induce trials or demonstrate the product at this stage to connect with and engage the consumer. Being candid is absolutely essential. Higher the level of engagement with the user, higher the 'pain', more will be the desire for possessing the product.

It is, however, essential to use the right triggers, based on the profile of the enlisted prospects. A campaign depicting high interest rates on term deposits will not engage a consumer in an Islamic country, where cultural taboos on accepting interest exist. Low bank charges or systematic investment plans could perhaps engage such a customer in a more productive way. Engaging industrial buyers involves more than just catch phrases and visuals. It is achieved using systematic cost-benefit analyses, comparative statements, simulations and sampling.

However, the core of this effort is to generate enough interest and understanding of your product or service that it prompts a consumer to believe that he will be better off after making a purchase. The promise of a better deal is hard to ignore.

Use of the product will prove its utility and cause feelings of satisfaction. The consumer feels **empowered**, having been provided a product or service that meets his specific needs. This empowerment temporarily curbs the 'pain' prompting the purchase, until the need is felt again. Repeat purchases make the consumer flaunt his "better deal" to others. The recognition, praise and satisfaction creates **enslavement** and attachment with the product and lays the foundation for brand loyalty. The user also talks about his 'conquest' to others. More prospects get enlisted and the cycle continues.

## The 'Pain-Pride' Relationship

Recall the first time you saw an advertisement for a portable computer that doubled up as a tablet and a notebook. As a consumer, this was the first time that you were exposed to the product and its benefits. Evaluating its features lead you to perceive the utility for it. A few days later, you are sitting cramped in a plane with the passenger in front having fully reclined his seat so you barely can move, let alone work on your laptop. A fellow passenger, seated next to you, is using the foldable machine you were considering to buy, instantly converting his laptop into a touch screen tablet, comfortably working while you stare.

The pangs of "Do I need it?" change to "I want it". The level of 'pain' peaks and the desire to possess manifests itself. A survey of available options follows naturally along with an evaluation of affordability. "Is it worth owning?" Purchase follows and 'pain' metamorphoses into 'pride'. As you use your new possession, pride grows and feelings of deprivation vanish.

Perhaps a year later, an image of a new brand of portable computers with superior features arrives on the billboards. 'Pain' resurfaces and marketers once again make you feel deprived. You tell yourself "I need more."

This cyclical process is the same whether we are buying a laptop or something as common as a T-shirt. Peaks and valleys of 'pain' appear with almost clockwork regularity. Every time pain peaks, we buy. If the product is found to be too expensive, the inner feeling of deprivation persists. It forces us to keep looking for alternatives, special promotions, direct or indirect substitutes. If the perceived need is intense, we find ways of borrowing to buy. The credit card industry works overtime to provide us with a way to spend money we have not earned to address two sources of 'pain' - the frustration of not having enough money to satisfy our needs and the desire of possessing the product itself.

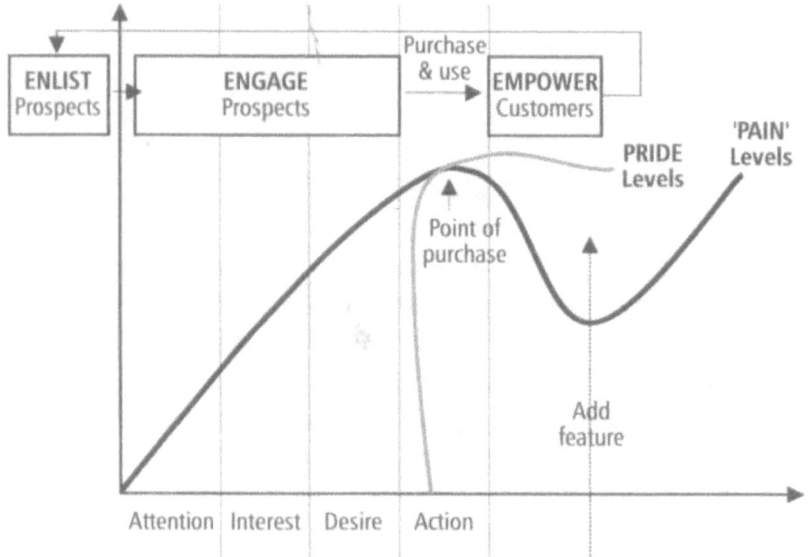

**Figure 3. The Pain-Pride Relationship shaping Consumer Intent**

### Pain, price and affordability

Awareness of every feature or benefit associated with the product or service serves as a catalyst in the 'pain' creation process within us. However, whether the feeling of deprivation is sustainable till a point that it leads to a purchase depends on several external and internal factors.

A consumer is repeatedly exposed to a product through advertising and marketing messages. He becomes aware of the features in the product and decides that it is a product that would be useful to him. However, upon making enquiries he realises that the price of the product is way above his budget. Though 'pre purchase pain' has risen, it may not lead to an eventual purchase.

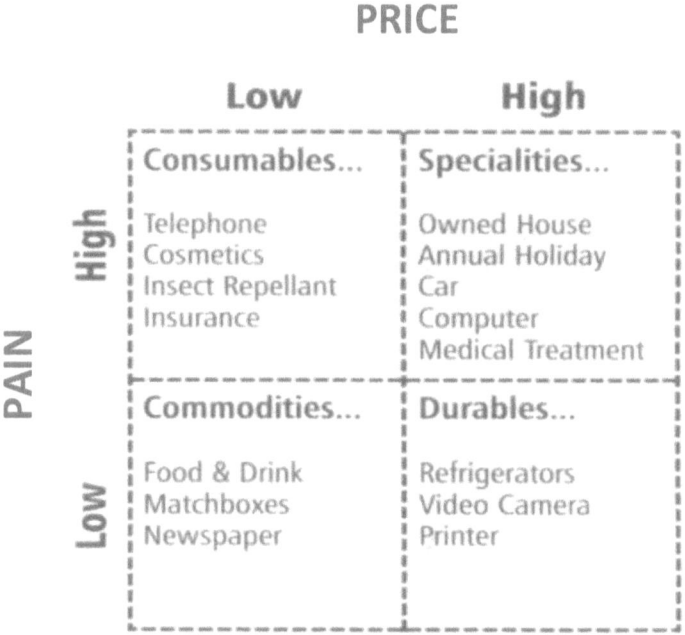

Figure 4. Levels of Pre-Purchase Pain

Levels of pain would also vary depending on the individual's preferences and prevailing circumstances. A person who has not eaten for a prolonged period would place even basic necessities of food in a 'high pain' segment. For a poor person who cannot afford even a square meal a day, food would be in a 'high pain - high price' quadrant.

Familiarity with the product is another factor. A person buying his second car or a third computer would have had a much higher inner feeling of deprivation when he had bought a similar product for the first time. This is true across all product categories. As we get used to a product, we take its utility for granted. But the need for "one more" certainly does cause enough pain within us leading us to buy an upgraded version of the same product.

Perhaps the most crucial determinant of the level of 'pain' within a prospective consumer is the timing of the decision. After a major natural calamity where an individual has lost a great deal, insurance products would suddenly become a top priority. Previously, the desire to purchase insurance policies may have been relegated to a much lower level. The need for a video camera peaks just before a vacation or a major family event.

Here, though all elements of the marketing mix remain constant, levels of pain rise. Seasonality may also cause differences in the classification of products. Looking at an exquisite pashmina shawl in the peak of the Indian summer would not cause enough 'pain' to induce an immediate purchase. A purchase may be prompted using inducements other than the product itself, most commonly a deep discount during the off season months to liquidate inventory. Make an offer that the consumer cannot afford to refuse or ignore.

### The Lure of a Better Deal

Marketers have always used their communications to make us realise that we need more or that their products and services are capable of giving us a better deal. The more they tell us, the more we filter and evaluate their brand messages. But marketing campaigns are persistent and our needs beyond any limit.

Levels of pain within each consumer would most certainly vary. For the same prospective buyer, the same product or service would evoke varying levels of desire at different points in time and within different environments. However, the inner feeling of deprivation undoubtedly manifests itself within us.

**It is this psychological 'pain' that serves as the primary influencer that drives us to buy.** We buy to curb this pain and buy more when the pain resurfaces. The satisfaction that arises post purchase is addictive. We do all we can to make the pain go away. We crave to have it all. And this is every marketer's dream.

# CHAPTER 3

# Enlisting Potential Buyers

The accurate identification of potential buyers or 'prospects' is the beginning of the sales process. We would not want to be targeting a set of consumers who do not feel adequately enough 'pain' of not owning or using our product. A family of three would not have enough of a reason to buy an eight room villa. However, a deal can successfully lure them to move from their one-bedroom studio apartment an hour's commute away from work to a spacious two-bedroom apartment located in a gated community fifteen minutes away from their workplace.

Creating such a list of potential buyers would involve studying habits of prospects and then segmenting them into groups using a structured 'Dart Board and Rubix Cube Method'.

**The Dart Board**

The Dart Board serves as an effective metaphor to represent a market and its segments. The bull's eye in the centre of the dart board represents the most sought after target market segment – one that every player aims to capture, but very few actually do. Insufficient practice, lack of patience, underdeveloped skills, poor vision, etc. are possible reasons why a player fails to capture the bull's eye. Each concentric circle around the bull's eye represent other market

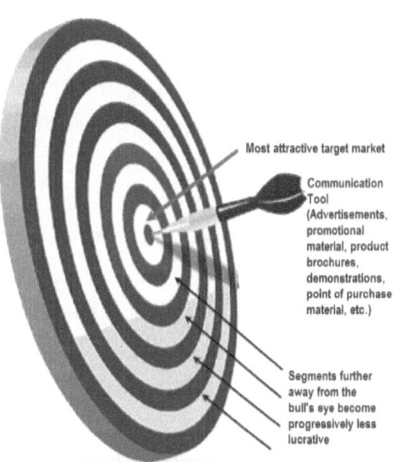

**Figure 5. The Dart Board for Market Segmentation**

segments, with each circle further away from the centre representing a progressively lesser attractive target market.

The dart represents the seller's positioning tool. Positioning is a battle for the consumer's mind and it is fought using appropriate communication methods. Sellers select what they believe is the most appropriate communication strategy and release the message (the 'dart') aimed at their intended target market. Once released, the seller has almost no further control of the message. Consumers will **react** to the message based on their own perception and understanding, **interpret** based on their own exposure and experience, as well as **retain** and **remember** based on their own personal choices. It is for this reason that even the most carefully crafted campaigns have highly variable success rates. Sometimes the message may also unintentionally appeal to a completely different market segment than originally designed for.

Applying this to the home buyer market, the bull's eye would perhaps be a location in the city centre, walking distance from the consumer's office. Though the most ideal home location, every buyer would not have the opportunity to live here, due to reasons ranging from affordability to limited availability. Hence the buyers would start looking at other market segments beyond the centre, each larger in size than the previous segment but further away from the city centre and correspondingly lower in price.

This segmentation is two dimensional. In case of the apartment purchase example, target segments are defined as 'who' will buy and 'where'.

## The Crucial Third Dimension

Adding a third dimension on the 'why people buy' can lead to several interesting outcomes. Consider the market for chocolates. Cadbury's have a range of several chocolate brands, some target the premium chocolate connoisseurs and have immense snob appeal, some target children for daily consumption, some bring out the romance in young adults. By adding the third 'why' dimension, it is now possible to understand the reasons 'why' a segment of affluent males aged 30 to 40 ('who') in Mumbai city ('where') buys a super-premium chocolate brand. They buy it to impress. Conspicuous consumption and visible display adds to their empowerment. Though taste is crucial, the branding adds an aura that goes beyond mere utility.

## The Market as a Rubix Cube

The market is often a puzzle, like a Rubix cube. To visually piece together a three dimensional market segmentation grid, a Rubix Cube provides a new viewpoint. The seemingly simple grid of the cube is deceptive. It is capable of moving in all directions and once jumbled, it can take great skill, patience, time and perseverance to unravel.

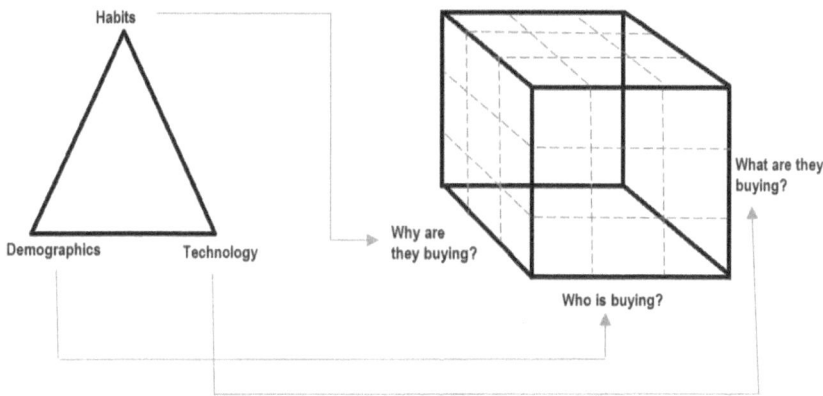

**Figure 6. Market Dynamics on a Rubix Cube**

The habits of consumers will decide their reasons to buy or preferences towards purchase of a particular product. A diabetic will also be a consumer for chocolates, because it is habitual for them to use it to correct bouts of hypoglycaemia, or sudden drop in blood sugar levels. They can also be a consumer for chocolates that are made with sugar substitutes. The chocolate brand that they will purchase in both these cases with however differ. To correct hypoglycaemia, they will use a regular chocolate brand that has high sugar levels. But in the case of a leisurely bite on a regular day, the brand will be a 'made for diabetics' product variant.

The illustration attempts to segment the chocolate market using three dimensions. The segment marked with the solid circle represents a *regular chocolate bar being used to correct low blood sugar in an urban area.* A diabetic could be a possible user in situations where she has had her

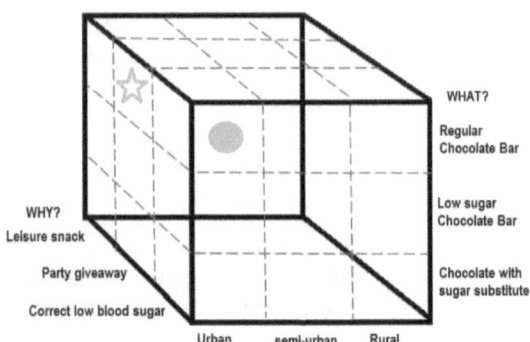

**Figure 7. Illustration of the Chocolates market on a Rubix Cube**

medication and not eaten a meal thereafter, leading to blood sugar levels dropping suddenly. Typically this is not a consumer who consumes a regular chocolate bar often and has restricted usage. Therefore, an ideal pack to meet this consumer's needs would be a mid-size bar with bite sized pieces sold in an easily re-sealable pouch that fits easily into the lady's purse for use once again at a later instance. This would be a product ideally designed for the buying needs of this consumer.

Now look at the segment marked with the star. It represents a *regular chocolate bar being purchased in an urban area as a party give away.* The same product will now need to be presented differently. Ideally it will now require to be packed in a 'party pack' with several units bundled together for easy distribution. Party planners expect to buy in bulk and usually at a discount. A packaging design variant with a bright, festive and fun look would be appreciated by this segment. Even if the product and consumer's location remain the same, the 'why people buy' changes the rest of the marketing mix.

Once all brands and variants are similarly plot on the cube, a marketer can now attempt to turn the cube's moving parts. This will shift the product into newer segments and could lead to the same product being positioned in completely new markets or to a completely new set of consumers.

**Unintended Consequences can unearth new market opportunities.**

Here are some examples of products that were used in unintended ways:

**Post-it Notes:** Post-it Notes were originally created as a low-tack, reusable adhesive to be used in bookmarks. However, they became popular for their unintended use as a way to leave temporary notes and reminders. People started using them to jot down quick messages and stick them to various surfaces, leading to their widespread use as a tool for organization and communication.

**Viagra:** Viagra was originally developed by Pfizer as a medication to treat high blood pressure and angina. However, during clinical trials, it was discovered that one of its side effects was its ability to induce erections in men. This unintended effect led to the repositioning of Viagra as a treatment for erectile dysfunction, and it has since become one of the most well-known and widely prescribed medications for this condition.

**Bubble Wrap:** Bubble Wrap was initially created as a textured wallpaper in 1957, but it failed to gain traction in that market. However, it was soon recognized for its unintended use as a protective packaging material. The air-filled bubbles provided excellent cushioning and protection for fragile items during shipping, leading to the widespread adoption of Bubble Wrap as a packaging material rather than a decorative wall covering.

In India, these examples illustrate how cultural and regional factors (why people buy?) can lead to the unintended use of products, often resulting in unique and unanticipated applications.

**Dettol as a Mosquito Repellent:** In many Indian households, Dettol, a popular antiseptic liquid manufactured by Reckitt Benckiser, has been used as a mosquito repellent. While Dettol is primarily intended for disinfecting wounds and surfaces, some people in India have used it by mixing it with water and applying it to their skin or using it in a vaporizer to repel mosquitoes. This unintended use of Dettol as a mosquito repellent is a common practice in some regions of India.

**Fair & Lovely Cream for Men:** Fair & Lovely, a skin-lightening cream marketed towards women, has been used by some men in India for its unintended purpose of lightening their skin tone. Despite being initially targeted at women, the product gained popularity among men seeking to achieve a fairer complexion. In response to criticism and changing societal attitudes, the brand has been rebranded as "Glow & Lovely" to be more inclusive of all genders, reflecting the unintended use and the evolving perceptions of skin tone in India.

These examples demonstrate how products can find entirely new and unexpected uses beyond their original intended purposes, expanding the market for the companies significantly.

# CHAPTER 4

# Design By Intent

Consumer intent must lie at the very core of the product design process. Understanding 'why' will assist in deciding 'what' features will appeal to the target audience.

The intent of every purchase is to solve a specific problem. Whether it is a woman purchasing a nail polish remover, a school student buying a dictionary, a teenager downloading an application on his mobile phone or a company buying a specific software for use by its accounts team, every time a consumer puts money on the table, their objective is to solve a problem.

Every feature included in the product design must have specific utility. A bank offers 'Priority Banking' services to its high net worth clients. The 'Priority Banking' product has been designed to suit specific needs like exclusivity, speed, personalisation and privacy. In return for high volume and high value of transactions carried out by such clients, bankers create a luxurious setting to make these clients feel special and valued. Opting to avail of these facilities comes at a cost that these clients choose to pay for. If the perceived value of these customized services is not commensurate to the add-on price charged, the service will find few takers.

## Decoding User Intent

To design products by intent, it is crucial to have a deep understanding of why users want or need a particular product. Product designers can then create solutions that address those specific intents.

**User intent can be categorized into three main types:**

**Functional Intent:** Functional intent refers to the practical purpose or task that the user wants to accomplish with a product. For example, a user may intend to buy a laptop for work-related tasks such as word processing, data analysis, or graphic design.

**Emotional Intent:** Emotional intent relates to the emotional experience or satisfaction that users seek from a product. This could include feelings of joy, comfort, security, or excitement. For instance, a user may intend to purchase a luxury watch not only for its functional features but also for the status and prestige it provides.

**Symbolic Intent:** Symbolic intent revolves around the symbolic meaning or representation associated with a product. Users may intend to use certain products as a way to express their identity, values, or affiliations. For example, someone may choose to buy an electric car to showcase their commitment to sustainability and environmental consciousness.

**Designing Products by Intent**

Designing products by intent requires a holistic approach that considers both functional and non-functional aspects. Here are some key considerations in the design process:

**User-Centered Design:** User-centered design places the user at the core of the design process. It involves conducting user research, creating personas, and involving users in the design feedback loop. By understanding user needs, desires, and pain points, designers can create products that align with user intent.

**Design Thinking:** Design thinking is a problem-solving approach that emphasizes empathy, ideation, prototyping, and iteration. It encourages designers to deeply understand user intent, challenge assumptions, and explore multiple solutions before settling on a final design. This iterative process helps ensure that the final product meets user expectations.

**Usability and User Experience (UX):** Usability and UX play a crucial role in designing products by intent. Products should be intuitive, easy to use, and provide a seamless experience that aligns with user intent. User testing and feedback loops are essential to identify usability issues and refine the design accordingly.

**Aesthetics and Emotional Appeal:** Products designed by intent also consider the emotional aspect of user experience. The aesthetics of a product can evoke specific emotions or create a connection with users. Colour palettes, typography, visual elements, and overall product design should align with the intended emotional response.

**Brand Identity and Values:** Designing products by intent also involves considering the brand identity and values associated with the product. The design should reflect the brand's personality, values, and positioning in the market. Consistency in design elements across different products helps strengthen brand recognition and loyalty.

**Sustainability:** In recent years, there has been an increased focus on sustainable design practices. Designing products by intent includes considering the environmental impact of the product throughout its lifecycle. This involves using eco-friendly materials, reducing waste, optimizing energy consumption, and promoting circular economy principles.

### Empowering Consumers by Satisfying Intent

The products that have given consumers exactly what they asked for are those which meet or even exceed consumer expectations.

### Amazon Echo

Launched in 2014, the Amazon Echo is a voice-controlled smart speaker that has transformed the way we interact with technology. Amazon's extensive research and development efforts, along with its deep understanding of consumer preferences and behavior, enabled the creation of a product that has been widely adopted by millions of consumers worldwide. The Echo has continued to evolve, with new models and features being introduced to enhance its capabilities and meet consumer demands.

**Tesla Model 3**

The Tesla Model 3, launched in 2017, is a fully electric vehicle that has gained widespread popularity due to its affordability, performance, and environmental benefits. Tesla's direct-to-consumer sales model and focus on customer feedback allowed them to create a product that met and exceeded consumer expectations. The Model 3 has been a commercial success, with Tesla consistently delivering vehicles to customers in a timely manner.

Having given consumers exactly what they asked for, these products have had a significant impact on their respective markets and have become household names, showcasing the importance of understanding and meeting consumer needs in the development process.

**Over-Engineering Fails**

Over-engineering refers to the process of designing a product or system with more capabilities and features than are actually needed. It almost always leads to a product that is overly complex, difficult to use, and ultimately fails in the market.

The Apple Newton, a personal digital assistant (PDA) released in 1993, was designed with advanced handwriting recognition software and a plethora of features, but it was plagued with issues such as slow performance, high cost, and an overly complicated user interface. These factors, along with the introduction of competing products like the Palm Pilot, contributed to the Newton's failure in the market.

The Segway PT, a two-wheeled electric vehicle introduced in 2001, was marketed as a revolutionary mode of transportation but was hindered by its high price, limited range, and a learning curve for users to master its operation. As a result, the Segway struggled to gain widespread adoption and faced stiff competition from more affordable and practical alternatives like bicycles and electric scooters.

Ford's EcoSport, launched in 2013, was an attempt to capture the burgeoning compact SUV market in India. However, the vehicle's high price point, starting at around ₹6.5 lakhs, made it difficult for Indian consumers to justify purchasing it over more affordable alternatives like the Maruti Suzuki Vitara Brezza and the Hyundai i20 Active. Moreover, the EcoSport's fuel efficiency was not up to par with its competitors, which further dampened its appeal in the Indian market.

The Nokia N9, released in 2011, was an innovative smartphone that featured a unique design and ran on the MeeGo operating system. However, Nokia's decision to switch to Windows Phone OS for its future smartphones left the N9 with limited app support and a short lifespan. This, coupled with its high price point of around ₹25,000, made it difficult for the N9 to gain traction in the Indian market.

### Simplicity Rules

Simple products often hold a unique place in the market due to their ease of use, affordability, and accessibility. These products have become category leaders by consistently delivering value to consumers and staying ahead of the competition.

Nespresso coffee machines have revolutionized the way people make coffee at home. With their compact design, user-friendly pod system, and high-quality coffee, Nespresso has become a category leader in the single-serve coffee market. The brand's commitment to sustainability, innovative technology, and a wide range of coffee blends has further solidified its position as a market leader.

Dyson's V6 Cord-free vacuum cleaner has been a game-changer in the vacuum cleaner market. Its lightweight design, powerful suction, and cord-free convenience have made it a popular choice among consumers. The V6 series has consistently remained a top seller in the cordless vacuum market, thanks to its innovative technology, performance, and ease of use.

Dove Soap is another example of a simple yet successful FMCG product. Dove's original beauty bar contains just a few basic ingredients such as sodium lauroyl isethionate, stearic acid, sodium tallowate or sodium palmitate, lauric acid, sodium isethionate, water, and sodium stearate.

What sets Dove apart is its unique marketing approach that focuses on promoting real beauty and body positivity. Their "Campaign for Real Beauty" challenged traditional beauty standards and resonated with consumers worldwide.

Oreo cookies have achieved immense success. The cookies consist of two chocolate wafers with a sweet cream filling in between. Oreo has become one of the best-selling cookie brands globally by offering a simple yet irresistible taste experience. The brand has also capitalized on nostalgia marketing by promoting the emotional connection people have with Oreo cookies and creating innovative variations and limited-edition flavours.

Pringles chips are known for their distinctive shape and packaging. These potato-based snacks are made from dried potatoes, vegetable oils, rice flour, wheat starch, and seasoning. Pringles' success can be attributed to their unique stackable shape, which not only stands out on store shelves but also offers a convenient snacking experience. The brand has also introduced a wide variety of flavours to cater to different consumer preferences.

Significant success of these products can be attributed to their simplicity. These products have resonated with consumers due to unique value propositions, convenience, and emotional appeal.

**The 'Intention to Try'**

The Theory of Trying developed by Bagozzi and Warshaw (1990) emphasises consumer uncertainty when achievement of a consumption objective is not entirely within one's volitional control. Impediments can take several forms: outcome uncertainty, lack of knowledge/information, distortion of market information, unfavourable earlier experiences, time pressure and cultural differences, need to be self-reliant and satisfaction with current behaviour, when new solutions require efforts in terms of search costs, transaction costs, etc.

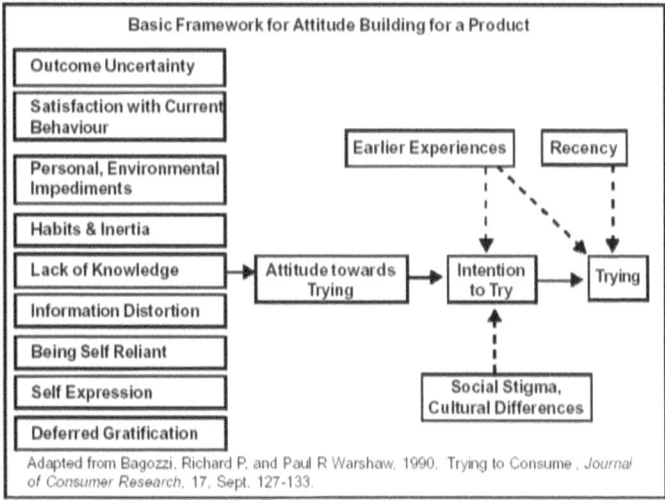

**Figure 8. Basic Framework for attitude building for a product**

Nine generic factors together affect the formation of an attitude towards trying a product. At times, even a favourable attitude towards trying the product, might not directly translate into an intention to try. Earlier experiences and socio-cultural norms applicable to the individual may also influence to some extent the intention to try.

After the consumer develops an 'intention to try', the next step is to actually try the product. However, besides the 'intention to try', actual trying is also affected by 'recency'. If the consumer has tried out a similar product in the recent past, he will be more amenable to trying out the product now. An important point to note is that earlier experiences affect both the 'intention to try' and 'actual trying'. When a person is not clear about his intentions to try out a product, he may rely upon his past experiences to decide whether he wants to actually try out the product.

It becomes essential therefore, while using intention surveys, to understand and appreciate the factors that affect the formation of an attitude towards the product / product category.

## Table 3. Factors Affecting the formation of
## Attitude towards a New Product

| No. | Factor | Consumer Thoughts |
|---|---|---|
| 1 | Personal and Environmental Impediments (price, availability, choice available) | • I can't afford this product<br>• The product is not easily available in the market<br>• I don't have time to go and look for new products |
| 2 | Outcome Uncertainty (availability of reliable evidence on utility and benefits) | • What if the usage of the new product does not demonstrate immediate effects?<br>• Is it effective in the long term?<br>• I have been advised to use an existing product to solve my problem. How can this new product help me better? |
| 3 | Satisfied with Current Behaviour | • I regularly use the existing product available in the market and that is enough to meet my needs<br>• I do not need anything new. |
| 4 | Habits and Inertia (resulting from laziness, indolence or as a coping mechanism in stressful situations. Eg., a person who knows that he is getting lower crop yields compared to others in the area, but denies it in an effort to escape reality.) | • I have never used any new product of this kind in the past. So why now?<br>• I don't want to use your product<br>• My forefathers have been using the existing product since generations. Why should I change the family's consumption trend? |

| 5 | Information Distortion | • I think these kind of new products are associated with specialized usage requirements only. I am not doing anything extraordinary to require their use.<br>• There have been some reports of your kind of products increasing cost of usage exorbitantly. |
|---|---|---|
| 6 | Deferred Gratification | • I have other priorities.<br>• I would rather spend on my family/ children. |
| 7 | Being Self Reliant | • I would rather modify the product I own currently to suit my needs, instead of buying something new. |
| 8 | Self Expression | • Using your new product is a passing fad.<br>• I don't believe in this. (Societal defiance) |
| 9 | Earlier Experiences | • Your product has short lasting effects – one needs to re-use very often<br>• I used something similar to your product once before. But I was not satisfied with the results. |

| 10 | Lack of Knowledge/ Information (concerns lack of awareness of the problem itself as well as of the product, price, availability etc, while 'Outcome Uncertainty' concerns effectiveness of the product) | • How does your product work to meet my needs?<br>• Are there any side effects?<br>• Will this increase my work or usage cycle time?<br>• I have been using similar products in the past but have not seen any noticeable results<br>• I have a specific disability. Can I use your product in these circumstances?<br>• Where do I get more information?<br>• Where is your product available in the market?<br>• What is the difference between existing products and your new product? |
|---|---|---|
| 11 | Social Norms and Cultural Differences | • What will other people say if they come to know? |

## From Understanding Intent to Forecasting Demand

Intentions lie at the core of a consumer's choices. These intentions are formed on the basis of product features, benefits, availability, simplicity, affordability, emotion, herd mentality, etc.

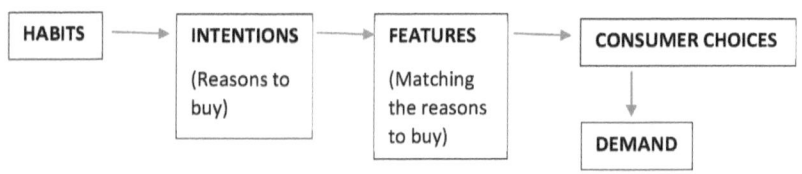

**Figure 9. From Habits to Intent to Demand**

If we ask a consumer for their intent, i.e. conduct a survey of reasons why they bought a product or service, we will get an insight into their mind. Using this description of their conscious choices will build a platform to understand the web of demand triggers.

# CHAPTER 5

# Forecasting The New: Evaluating Methods & Models

Forecasting demand for new products and in turbulent markets requires us to stick our neck out without a reliance on past data sets or established norms. We do not have the possibility of standing on the broad shoulders of past trends and the ease of extrapolation.

Past data, especially in the current post Pandemic, highly complex and dynamic world, cannot be relied upon to make forecasts. Certainty, longer product life cycles and low competitive intensity are things of the past. Increased consumer awareness, understanding and acceptance for new products and processes have set the industry's "ground" itself in motion creating an environment that can be aptly classified as a "turbulent field". Demand has become uncertain, product life cycles have shortened, and competition has intensified. In such a situation, understanding demand, planning demand and linking supply with demand is crucial.

'The New Future' does not look like a repeat or extension of the past. Consumer Habits have evolved and technology has found new solutions that have reconfigured the very foundations of intent. Hence there is a need to take a hard look at how we can map the minds of the consumers and make reasonably accurate demand forecasts in the absence of past data.

The rapidly evolving environment is the source of constraints, contingencies, problems, and opportunities.

Forecasting demand in such dynamic industry environments is a challenge. This is accentuated for companies marketing specialty products or new product concepts where past demand data is unavailable, and the web of demand triggers is known but constantly changing.

It becomes essential to comprehend the individual impact of each factor affecting demand separately. Strategies may be formulated to 'influence the influencers' based on this information. Demand estimates would also assist in devising operational frameworks, planning inventories, increasing manufacturing and supply chain efficiency, etc.

## The Web of Demand Triggers

The forecasting process begins with an understanding of what triggers demand. It is possible to disaggregate (or decompose) the causal factors that lead to a consumer deciding to purchase a product or service.

Contrary to the common assumption, time series for demand are not always subject to consistent forces that point in the same direction. Changes in the trend in a time series arise due to a variety of underlying factors. The term 'causal forces' is used to represent the cumulative directional effects of the factors that influence the trends in a time series. Causal forces are classified by the way that they relate to historical trends. Some forces cause movement in a given direction (up or down) irrespective of current trends; called 'growth' and 'decay' forces. Forces that support the current direction of a trend are called 'supporting'. Forces that act against trends are labelled as 'opposing.' Finally, 'regression' is used to classify forces that cause a series to move toward a mean.

Forecasting complex time series (like demand forecasts for specialized products) would be easier if one could decompose the series to eliminate the effects of the conflicting forces. In a study by Yokum, Collopy and Armstrong (2004), for nine series post decomposition of causal factors, the forecast error was reduced by more than half. For three series, decomposition by causal forces had little effect on accuracy.

MacGregor (2001) noted that decomposition is expected to be more valuable in situations involving high uncertainty. Uncertainty is a core feature of today's turbulent business environments, and therefore decomposition to identify individual causal factors becomes a forecasting imperative. Furthermore, the benefit of using causal forces increases as the forecast horizon lengthens because the causal effects increase accordingly.

Considering the limited success of using complex econometric methods in forecasting for new industries or specialized products and the unavailability of reliable, published past time series of the demand data for many industries, it may not be prudent to try and apply pure econometric modelling for forecasting demand. However, the use of intention surveys which have improved forecasting accuracy in several fields holds promise. Prior to using such an intentions survey, considering the high level of uncertainty which characterizes today's markets, it would be useful to decompose the causal factors that influence the overall demand for a particular product or service and then attempt to develop the forecasting model.

## Overview of Traditional Demand Forecasting Techniques

Significant gains have been made in forecasting for marketing in the past few decades. Advances have occurred in the development of qualitative methods such as Delphi, role playing, intentions and opinions surveys, and bootstrapping. They have also occurred for quantitative methods such as extrapolation and econometrics. An attempt is made in this section to build on the experience in applying these methods by researchers so generalizations can be made about which methods would be most appropriate to forecast demand for turbulent markets.

In general, experts advocate the use of structured methods that avoid intuition, unstructured meetings, focus groups, and data mining. In situations where there is sufficient data, use of quantitative methods is encouraged, including extrapolation, quantitative analogies, rule-based forecasting and causal methods. In other cases, it is appropriate to use methods that structure judgement including surveys of intentions and expectations, judgmental bootstrapping, structured analogies and simulated interaction. Green & Armstrong (2005) strongly advocate the integration of Judgmental and statistical methods. Domain knowledge should be incorporated into statistical forecasts.

## Econometric methods

"Econometric methods" are defined as quantitative approaches that attempt to use causal relationships in forecasting. In particular, they refer to models based on regression analysis and include all methods which forecast by explicitly measuring relationships between the dependent variable and some causal variables. (Armstrong, 1978)

For market demand forecasting, there is empirical evidence to support the use of econometric methods rather than subjective methods for long-range forecasts. To date, most econometric researchers have devoted their efforts to short-term forecasting, an area that has yielded unimpressive or contradictory results (Armstrong, 1985; Fildes, 1985). Econometric methods would be expected to be more useful for long-range forecasting because the changes in the causal variables are not swamped by random error, as in the short run. In fact, econometric methods are more accurate. Armstrong (1985) reported seven empirical comparisons of methods used in long-range forecasting. In all comparisons econometric methods were more accurate than extrapolations. Also, there was a 3 to 0 advantage for econometric versus subjective forecasts. Fildes (1985) located 20 studies on long-range forecasting; he coded them as 15 showing econometric to be more accurate, 3 ties, and 2 showing econometric to be less accurate than other methods.

Thus it may be concluded that Causal econometric methods provide more accurate long-range forecasts. While more expensive, the methods are expected to be the most accurate method when large changes are expected (like in today's turbulent markets). What must however be borne in mind is that to improve predictive capacity, these causal models need not be complex. (Armstrong, Brodie & McIntyre, 1987)

### Naive versus Causal Methods

A continuum of causality exists in forecasting models. At the naive end, no statements are made about causality (sales can be plotted against time and the trend can be projected); at the causal end, the model may include many factors (the real income per capita, inflation, the literacy levels in the population, and the real price of substitute forms of the product).

Causal methods are more complex than naive methods. First, data must be obtained on the causal factors. Estimates of causal relationships are obtained from these data. These estimates of the causal relationships should be, adjusted so that they are relevant over the forecast horizon. Next, one must forecast the changes in the causal variables. Finally, the forecasts of the causal variables and the relationships are used to calculate the overall forecast.

Causal methods are of more obvious value in forecasting. However, naive methods can be used in some phases. For example, naive methods can provide forecasts of environmental factors.

**Intentions Surveys**

Theoretical literature in psychology that suggests that a good predictor of an individual's future behaviour is his or her stated intention (Fishbein and Ajzen 1975). However, the psychological literature also suggests that past behaviour is an important predictor of future behaviour (Bentler and Spechart 1979).

With intentions surveys, people are asked how they intend to behave in specified situations. In a similar manner, an expectations survey asks people how they expect to behave. Expectations differ from intentions because people realize that unintended things happen. For example, if you were asked whether you intended to purchase a product in the next month you might say no. However, you realize that a problem might arise that would necessitate such a purchase, so your expectations would be that the event had a probability greater than zero. This distinction was proposed and tested by Juster (1966) and its evidence on its importance was summarised by Morwitz (2001). (Green & Armstrong, 2005)

Expectations and intentions can be obtained using probability scales such as Juster's eleven-point scale. The scale should have descriptions such as 0 = 'No chance, or almost no chance (1 in 100)' to 10 = 'Certain, or practically certain (99 in 100)'. (Green & Armstrong, 2005)

To forecast demand using a survey of potential consumers, the administrator should prepare an accurate and comprehensive description of the product, its benefits and conditions of sale. He should select a representative sample of the population of interest and develop questions to elicit expectations from respondents. (Green & Armstrong, 2005)

Purchase intentions are routinely used to forecast demand of existing products and services. While past studies have shown that intentions are predictive of sales, they have only examined the absolute accuracy of intentions, not their accuracy relative to other forecasting methods. (Kumar, Morwitz & Armstrong, 2000)

For different products and time horizons, intentions-based forecasting methods were more accurate than an extrapolation of past sales. Combinations of these forecasting methods using equal weights lead to even greater accuracy, with error rates about one-third lower than extrapolations of past sales. Thus, it appears that purchase intentions can provide better forecasts than a simple extrapolation of past sales trends. (Kumar, Morwitz & Armstrong, 2000)

Purchase intentions are inexpensive to acquire and easily understood, which may account for their widespread use. (Kumar, Morwitz & Armstrong, 2000)

Many studies have found a positive correlation between purchase intentions and purchase behavior (Adams 1974; Juster 1966; McNeil 1974; McNeil and Stoterau 1967; Morwitz and Schmittlein 1992; Morwitz, Steckel and Gupta 1996; Tobin 1959, all cited in Kumar, Morwitz & Armstrong, 2000).

Buyer-Intentions and expectations surveys are especially useful in forecasting demand when demand data are not available, such as for new product forecasts. (Silk and Urban 1978, Green & Armstrong, 2005)

The theoretical literature is equivocal about whether intentions-based forecasts or past sales trends should be more accurate. Received wisdom suggests that the best predictor of future behaviour is past behaviour. On the other hand, the social psychology literature states that a good predictor of what individuals will do is their stated intentions to perform the behaviour (Fishbein and Ajzen 1975).

Other research suggests that intentions data are useful for predictions under certain conditions. Armstrong (1985) summarizes these conditions: (1) the event being predicted is important, (2) the respondent has a plan (at least the high intenders do), (3) the respondent can fulfil the plan, (4) new information is unlikely to change the plan over the forecast horizon, (5) responses can be obtained from the decision maker, and (6) the respondent reports correctly.

Such conditions are likely to be met for purchase intentions of "high involvement" goods and services, like buying a house, a motor vehicle, a life insurance policy, a gym membership, specialized farm technology, etc. Once convinced of the utility, the consumer makes a definite plan to buy the product or service at a fixed time. With the increase in purchasing power, the consumer is in a position to convert potential demand into effective demand. It is also possible to accurately obtain intentions information from consumers. This suggests that intentions data could potentially improve accuracy of forecasts based solely on past sales behaviour for these products. (Kumar, Morwitz & Armstrong, 2000)

Another reason why intentions surveys will be useful to predict demand in certain markets is the fact that past and current demand data is not available from any authentic source. The nascent industry has a large number of unorganized players whose sales information is not recorded or reported in formal databases.

A variety of survey questions have been used to measure consumers purchase intentions. Among the most commonly used measures are Juster's 11-point purchase probability scale (Juster 1966) and a 5-point likelihood of purchase scale. Day, et al. (1991) concluded that Juster's 11-point purchase probability scale provides substantially better predictions of purchase behaviour than other types of intentions scales.

However, considering that the respondents in would be more at ease with indicating purchase probabilities on a simpler user friendly scale, it may be preferable to use a shorter purchase probability scale.

Purchase probabilities (e.g., What is the probability that you will buy life insurance?) and expectations (e.g., How likely are you to buy this life insurance policy? Do you expect to buy a life insurance policy?) are broader than direct intentions questions (e.g., Do you intend to buy a life insurance policy? Do you plan to buy life insurance?) because they refer to actions that might be unplanned as well as planned.

Assessing purchase probabilities and expectations may be advantageous in situations where people realize that they may purchase an item even though they have no plans at the time of the survey (e.g., they realize while answering the research questionnaire that their next crop would have higher, unique nutritional needs that may necessitate the changeover to more specialized fertilizers). Therefore, a smaller proportion of respondents report "zero" on purchase probability scales than report "no" on intentions scales (Juster 1966).

For many studies, most purchases are made by those who had reported no plans to buy (Juster 1966). This occurs because although non-intenders seldom purchase, they are often the largest group of respondents. Models have been developed to describe how purchase intentions relate to purchase behaviour.

Two commonly used methods to forecast sales from intentions predict that the proportion of consumers who will purchase will equal:

(1) the mean intent (transformed to lie between zero and one to represent the mean probability of purchase), or (2) the proportion of respondents indicating a positive purchase intent (Morwitz et al. 1996).

Intentions, by themselves, provide only a crude way to predict sales. Several studies have shown that these methods often provide biased estimates of sales, overstating or understating actual purchasing (Morrison 1979, Manski 1990). Thus, when possible, sales data should be used to adjust for the bias in intentions. The simplest way to do this is to relate an aggregate measure of purchase intentions to an aggregate measure of sales.

For new products or new product groups (e.g. buying a satellite phone in an area or by a potential consumer who has never been exposed to the concept), intentions are sometimes used directly to forecast demand. However, when sales figures are available, it is sensible to calibrate intentions against them.

In other words, we look at a category of intenders (e.g., "we are certain we will buy a satellite phone in the next three months") and determine what percent actually did purchase in that period. This relationship is then extended to the period to be forecast. (Kumar, Morwitz and Armstrong, 2000)

Morrison (1979) developed a descriptive model of the relationship between purchase intentions and subsequent purchasing. Morrison proposed that there are three threats to the predictive validity of purchase intention measures. First, intentions are measured with error. Second, respondents' purchase intentions might change over time because of exogenous events (e.g., current period is affected by an adverse business cycle, sudden rise in income, unfavourable weather conditions). Third, average stated purchase intentions might be a biased estimate of the proportion that actually buy the product because of systematic error (e.g., response style biases, sales promotional effects, changes in the economy, as noted in Kalwani and Silk 1982).

However, it was uniformly observed that for different products, time horizons, countries, and types of intentions questions, the intentions surveys when combined with prior sales data, were more accurate than forecasts based solely on past sales.

### Delphi Technique

Since its design at the RAND Corporation over 40 years ago, the Delphi technique has become a widely used tool for measuring and aiding forecasting (Rowe and Wright, 1999)

Delphi is not a procedure intended to challenge statistical or model-based procedures. It is intended for use in judgment and forecasting situations in which pure model-based statistical methods are not practical or possible because of the lack of appropriate historical /economic/ technical data, and thus some form of human judgmental input is necessary (e.g., Wright, Lawrence & Collopy, 1996). Such input needs to be used as efficiently as possible, and for this purpose Delphi technique might serve a role. (Rowe and Wright, 1999)

Four key features may be regarded as necessary for defining a procedure as a 'Delphi'. These are: rounds. anonymity, iteration, controlled feedback, and the statistical aggregation of group response. (Rowe and Wright, 1999)

Anonymity is achieved through the use of questionnaires. By allowing the individual group members the opportunity to express their opinions and judgments privately, undue social pressures – as from dominant or dogmatic individuals, or from a majority – should be avoided. Ideally, this should allow the individual group members to consider each idea on the basis merit alone, rather than on the basis of potentially invalid criteria (such as the status of an idea's proponent). (Rowe and Wright, 1999)

Furthermore, with the iteration of the questionnaire over a number of rounds, the individuals are given the opportunity to change their opinions and judgments without fear of losing face in the eyes of the (anonymous) others in the group. Between each questionnaire iteration, controlled feedback is provided through which the group members are informed of the opinions of their anonymous colleagues. (Rowe and Wright, 1999)

The number of rounds is variable, though it seldom goes beyond one or two iterations (during which time most change in panelists' responses generally occurs) (Rowe and Wright, 1999)

To forecast with Delphi the administrator should recruit between five and twenty suitable experts and poll them for their forecasts and reasons. The administrator then provides the experts with anonymous summary statistics on the forecasts, and experts' reasons for their forecasts. The process is repeated until there is little change in forecasts between rounds – two or three rounds are usually sufficient. The Delphi forecast is the median or mode of the experts' final forecasts. Software to guide you through the procedure is available. Rowe and Wright (2001) provide evidence on the accuracy of Delphi forecasts. The forecasts from Delphi groups are substantially more accurate than forecasts from unaided judgment and traditional groups, and are somewhat more accurate than combined forecasts from unaided judgment. (Green & Armstrong, 2005)

The use of questionnaires which are to be filled in without the presence of the researcher while using the Delphi technique may not be practically possible in the rural scenario. Language barriers may hamper response rates and response accuracy. Communicating with the respondent using the mail system may also cause unnecessary delays and constrain progress.

## Unaided Judgement

It is common practice to ask experts what will happen. This is a good procedure to use when experts are unbiased, large changes are unlikely, relationships are well understood by experts (e.g., demand goes up when prices go down), experts possess privileged information and experts receive accurate and well-summarized feedback about their forecasts. Unfortunately, unaided judgement is often used when the above conditions do not hold. Green and Armstrong (2005), for example, found that experts were no better than chance when they use their unaided judgement to prepare forecasts in complex situations.

## Game Theory and Role Playing

Game theory has been touted in textbooks and research papers as a way to obtain better forecasts in situations involving negotiations or other conflicts. Despite a vast research effort, there is no research that directly tests the forecasting ability of game theory. However, Green (2002, 2005) tested the ability of game theorists, who were urged to use game theory in predicting the outcome of eight real (but disguised) situations. In that study, game theorists were no more accurate than university students. (Green & Armstrong, 2005)

Role playing is well-suited to forecasting how people will respond to exogenous pressures (actions of those outside the firm). (Armstrong, Brodie & McIntyre, 1987) Such exogenous events (e.g. weather, government policy, etc) are inherent in the turbulent markets, and hence role playing may be considered as a useful tool in the forecasting process.

The accuracy gain of game theory over unaided judgment may be illusory, and the advantage of role playing over game theory is likely to be greater than the error reduction found by Green. The improved accuracy of role playing over game theory was consistent across situations. For those cases that simulated interactions among people with conflicting roles, game theory was no better than chance, whereas role-playing was more accurate. (Armstrong, 2002)

**Bootstrapping**

According to Armstrong, Brodie & McIntyre (1987) bootstrapping (including related approaches such as expert systems and conjoint analysis) is one of the more important advances for forecasting in marketing over the past quarter century. It was also noted as one of the most significant advances in the field of agricultural forecasting. (Armstrong, 1994)

Bootstrapping has been widely applied in marketing. Occasionally it has been used with experts, but typically it is consumer intentions that are modeled. Over 1,000 marketing applications had been made by indirect bootstrapping of consumer intentions by the early 1980s (Cattin and Wittink, 1982). These applications have been done under the umbrella term "conjoint analysis." Bootstrapping is nearly always more accurate than judgment (Camerer, 1981)

**Focus Groups**

One popular type of survey, focus groups, violates five important principles and they should not, therefore, be used in forecasting. First, focus groups are seldom representative of the population of interest. Second, the responses of each participant are influenced by the expressed opinions of others in the group. Third, a focus group is a small sample – samples for intentions or expectations surveys typically include several hundred people whereas a focus group will consist of between six and ten individuals. Fourth, questions for the participants are generally not well structured. And fifth, summaries of focus groups responses are often subject to bias. There is no evidence to show that focus groups provide useful forecasts. (Green & Armstrong, 2005)

However, in order to understand the context or the dynamics of the market environment, discussions using focus groups are useful. They may not be useful in specifically developing a forecasting model, but can certainly be used in understanding the context of a consumer's decision making process.

**Neural Nets**

Neural networks are computer intensive methods that use decision processes analogous to those of the human brain. Like the brain, they have the capability of learning as patterns change and updating their parameter estimates. However, much data is needed in order to estimate neural network models and to reduce the risk of over-fitting the data (Adya and Collopy 1998). There is some evidence that neural network models can produce forecasts that are more accurate than those from other methods (Adya and Collopy 1998). While this is encouraging, the current advice of experts like Green & Armstrong (2005) is to avoid neural networks because the method ignores prior knowledge and because the results are difficult to understand.

**Data Mining**

Data mining ignores theory and prior knowledge in a search for patterns. Despite ambitious claims and much research effort, we are not aware of evidence that data mining techniques provide benefits for forecasting. In their extensive search and reanalysis of data from published research, Keogh and Kasetty (2002) found little evidence for that data mining is useful. (Green & Armstrong, 2005)

**Segmentation**

Segmentation involves breaking a problem down into independent parts, using data for each part to make a forecast, and then combining the parts. For example, we could forecast sales separately for each climatic region, and then add the forecasts.

To forecast using segmentation, one must first identify important causal variables that can be used to define the segments, and their priorities. For example, nature of crop and proximity to assured irrigation facilities are both likely to influence demand for specialized farm technology, but the latter variable should have the higher priority; therefore, segment by irrigation, then crop. For each variable, cut-points are determined such that the stronger the relationship with dependent variable, the greater the non-linearity in the relationship, and the more data that are available the more cut-points should be used. Forecasts are made for the population of each segment and the behaviour of the population within the segment using the best method or methods given the information available. Population and behaviour forecasts are combined for each segment and the segment forecasts summed.

Where there is interaction between variables, the effect of variables on demand are non-linear, and the effects of some variables can dominate others, segmentation has advantages over regression analysis (Armstrong 1985).

Efforts at dependent segmentation have gone under the names of microsimulation, world dynamics, and system dynamics. While the simulation approach seems reasonable, the models are complex and hence there are many opportunities for judgemental errors and biases. Armstrong (1985) found no evidence that these simulation approaches provide valid forecasts and there appears no reason to change this assessment. (Green & Armstrong, 2005)

**Rule Based Forecasting**

Rule-based forecasting (Collopy and Armstrong, 1989) incorporates information from experts and from prior research. The procedure calls for the development of empirically validated and fully disclosed rules for the selection and combination of methods.

When large changes are expected, one should draw upon methods that incorporate causal reasoning. If the anticipated changes are unusual, judgmental methods such as Delphi would be appropriate. If the changes are expected to be large, the causes are well understood, and if one lacks historical data, then judgmental bootstrapping can be used to improve forecasting. (Green & Armstrong, 2005)

**Suitability of Forecasting Methods for Turbulent Markets**

The foregoing review of various forecasting methods and the reasons for the decision to use or reject the use of a particular method for forecasting in turbulent markets is summarized in the table below:

### Table 4. Evaluation and Rationale for Choice of Forecasting Techniques

| *Forecasting Technique* | *Decision regarding usage for a (near term) demand forecasting model for turbulent & new markets* | *Rationale for decision (for details and sources, please see the relevant sections above)* |
|---|---|---|
| Econometric Methods | **LIMITED USE** | Primarily useful for long range forecasts; unimpressive results in short range forecasting; past demand data unavailable for extrapolation |
| Naïve Methods | **LIMITED USE** | No assessment made about causality of specific factors |
| Causal Methods | **METHOD MAY BE USED** | Method used must lead to decomposition of 'causal forces' and assessment of impact of each causal factor on the demand for the new of specialty product, thus improving forecasting accuracy. |

| Intentions Surveys | **METHOD MAY BE USED** | Intentions based forecasts more accurate than extrapolation of past sales; especially useful when past demand data not available; collecting 'probability of purchase' estimates from potential consumers is not complicated |
|---|---|---|
| Delphi Technique | **LIMITED USE** | Administering Delphi questionnaires anonymously and without the researcher being physically present in a new, uninformed or rural setting would be a constraint. |
| Unaided Judgement | **LIMITED USE** | Experts were no better than chance while developing forecasts in complex situations. |
| Game Theory | **LIMITED USE** | No research has directly tested the forecasting accuracy of game theory |
| Role Playing | **LIMITED USE** | Useful only in understanding how people respond to exogenous pressures (government policy, weather, etc.) |
| Focus Groups & In depth interviews | **METHOD MAY BE USED** | Useful only for understanding the decision context and trends. Unstructured discussions rarely lead to accurate forecasts. However, they are of value in understanding market dynamics |

| Neural Nets | **LIMITED USE** | High data requirements; results generated using computer intensive methods are often difficult to interpret & understand |
|---|---|---|
| Data Mining | **LIMITED USE** | Little evidence of utility in forecasting |
| Segmentation Method | **METHOD MAY BE USED** | Increases depth and improves accuracy of the overall forecast. The intentions survey will gather data based on segments. Thereafter, demand estimates will be made for each segment separately and then summed to generate the overall forecast. |
| Rule Based Forecasting | **LIMITED USE** | Absence of empirically validated, fully disclosed prior rules that can be applied to the forecasts. |

## Other Forecasting Imperatives

Once the accurate method is identified to build a forecast, there are some other areas that deserve attention:

### Identifying causal variables

Environmental forecasts are useful as an input to strategic planning. The identification of possible states of the environment and a forecast of their likelihood can provide ideas on future demand trajectories. Environmental forecasts also can help to provide better industry forecasts (e.g. the total demand for a product class in a given market).

It is important that the forecasting methods first identify the possible states of the future. For this, brainstorming among a variety of experts would be useful. Particular attention would be given to the more important of these possible states. Importance should be judged not only by the likelihood of the environmental change, but also by its potential impact on the Industry if it does occur. For example, the interlinking of India's Rivers is a mega project that will most certainly affect the farm inputs industry and the rest of the agribusiness sector. It becomes important while forecasting demand to assess the likelihood of this event occurring and therefore, its potential impact on altering demand patterns of the industry.

There is some evidence to show that the accuracy of forecasts of environmental variables is not as important as is identifying the key variables to include in the market forecasting model. (Armstrong, Brodie & McIntyre, 1987). Measurement error in the causal variables (e.g., the environmental inputs to a market forecasting model) had little impact on the accuracy of an econometric model. (Denton and Kuiper, 1965; Denton and Oksanen, 1972 and McDonald, 1975).

It is important to determine the important factors in the environment that might affect the industry. It is also important to predict the direction of change in the important factors, and to then get "approximately correct" predictions of the magnitude of the impact of changes in these factors. (Armstrong J, cited in Kenneth Albert, 1983)

For the direction of change in environmental factors, only general trends, not cycles, should be considered. Other than recurrent events owing to the seasons of the year (seasonality), cycles have been of little value for improving the accuracy of forecasts. The reason? One must also predict the phases (timing) of the cycles. If the timing is off, large errors can occur. Organizations should have a system for scanning the environment to be sure that they do not overlook variables that may have a large impact on their market. These variables can be tracked through marketing information systems. Periodic brainstorming with a heterogeneous group of experts should be sufficient to identify which variables to track. The key is to identify the important variables and the direction of their effects. Once identified, only crude estimates of the coefficients of these variables are typically sufficient in order to obtain useful forecasts. When large shocks are encountered, more sophisticated approaches may be useful. (Armstrong, Brodie & McIntyre, 1987)

**Estimating Uncertainty**

In addition to improving accuracy, forecasting is concerned with assessing uncertainty. This can help manage the risk associated with the forecasts. Much work has been done on judgmental estimates of uncertainty. One of the key findings is that judges are typically overconfident. Fischoff and MacGregor (1982) found that 95% confidence ranges that are estimated judgmentally typically fail to include the true value. This bias occurs even when subjects are warned in advance about the overconfidence phenomenon. Nevertheless, judgmental expressions of uncertainty have been found to be useful.

One way to assess uncertainty has been to examine the agreement among judgmental forecasts. For example, Ashton (1985), found that the agreement among the individual judgmental forecasts was a useful proxy for accuracy.

Probably the best way to assess uncertainty is to follow the track record of a given forecasting method in actual use. (Armstrong, Brodie & McIntyre, 1987)

Traditional error measures, such as the mean square error (MSE), do not provide a reliable basis for comparison of forecasting methods (Armstrong and Collopy, 1992). The median absolute percentage error (MdAPE) is more appropriate because it is invariant to scale and is not influenced by outliers. When comparing methods, especially when testing on a small number of series, control for degree of difficulty in forecasting by using the median relative absolute error (MdRAE), which compares the error for a given model against errors for the naive, no change forecast (Armstrong and Collopy, 1992).

Statisticians have relied heavily on tests of statistical significance for assessing uncertainty. However, statistical significance is inappropriate for assessing uncertainty in forecasting. Furthermore, its use has been attacked as being misleading (e.g., see Cohen, 1994). It is difficult to find studies in marketing forecasting where statistical significance has made an important contribution.

Instead of statistical significance, the focus should be on prediction intervals. Chatfield (2001) summarizes research on prediction intervals. Unfortunately, prediction intervals are not widely used in practice. Tull's (1967) survey noted that only 25% of 16 respondent companies said they provided confidence intervals with their forecasts. Dalrymple (1987) found that 48% did not use confidence intervals, and only 10% 'usually' used them. (Green & Armstrong, 2005)

In a survey of experts by Yokum and Armstrong (1995) half said that it was important 'that your forecasting methods provide confidence bounds on the forecasts', while 20% said this was not important.

**Overlooked Discontinuities**

Considering the wide range of random shocks that affect markets, there is strong agreement about the importance of discontinuities in forecasting. This was surprising because this topic has been largely ignored in the forecasting literature. (Callopy and Armstrong, 1992)

Identifying areas of uncertainty or disagreement among experts, or disagreements between researchers and practitioners, could help to guide further research. Also, the opinions might aid in the development of expert systems for forecasting.

In a study of experts by Callopy and Armstrong (1992), 92% of the experts agreed that "abrupt changes" are an important consideration while forecasting demand. This is surprising given that time series forecasting research and practice have largely ignored abrupt changes. Examination of a convenience sample of indices of 28 books that discuss time series forecasting did not include any reference to 'abrupt changes', 'discontinuities', 'erratic fluctuations', 'interruptions', 'irregularities', 'ramps', 'shifts', 'steps', and variations on these terms. (Callopy and Armstrong, 1992) The experts agreed that seasonality and recent trend were key features. The experts also placed a heavy emphasis on the importance of abrupt changes in the historical data patterns. This stands in stark contrast to forecasting methods and forecasting research which have long ignored abrupt changes. We have no explanation for this mystery of the overlooked discontinuities. Fortunately, software developers are responding to this problem.

## Combining Forecasts

Considerable literature has accumulated over the years regarding the combination of forecasts. The primary conclusion of this line of research is that forecast accuracy can be substantially improved through the combination of multiple individual forecasts.

Clemen (1989) is a milestone on the topic of combining forecasts. As noted by Clemen, past research has produced two primary conclusions, one expected and one surprising. The expected conclusion is that combined forecasts reduce error (in comparison with the average error of the component forecasts). The unexpected conclusion is that the simple average performs as well as more sophisticated statistical approaches.

Combining forecasts is more useful for long-range forecasting because of the greater uncertainty. (Armstrong, 1989) The level of aggregation of the data was expected to be related to the relative accuracy of alternative extrapolation methods by 88% of the experts. We speculate that the level of aggregation may be important because different causal factors might affect different components. (Callopy and Armstrong, 1992) Highly aggregated data are more likely to be subject to different causal factors than are less aggregated data. On the other hand, the reliability of data often improves when one uses larger aggregates. 83% of the experts with an opinion believe that combining will produce more accurate forecasts.

Clemen (1989) advises forecasters to select a set of methods that differ substantially from one another with respect to the data used and also with respect to the procedures for analyzing the data. The experts believed that, in general, combined forecasts are more accurate than those based on a single method: 73% of the respondents agreed and only 15% disagreed.

Combined forecasts improve accuracy and reduce the likelihood of large errors. In a meta-analysis (the 'study of studies': Armstrong, 1989), Armstrong found an average error reduction of about 12% across 30 comparisons. They are especially useful when the component methods differ substantially from one another. For example, Blattberg and Hoch (1990) obtained improved sales forecast by averaging managers' judgmental forecasts and forecasts from a quantitative model. Considerable research suggests that, lacking well-structured domain knowledge, unweighted averages are typically as accurate as other weighting schemes (Armstrong, 2001).

Callopy and Armstrong (1992) favoured simple methods of preparing and combining forecasts for stable and unstable situations, with a slightly stronger preference for their use in unstable situations. Schnaars' (1986) results implied that simple models are most appropriate for unstable situations. The use of a simple average has proven to do as well as more sophisticated approaches. An alternative simple approach, the median, might offer additional benefits. It is less likely to be affected by errors in the data. Whether the median is superior to the mean is an empirical issue. Meta-analysis may prove useful here. Two studies that address this issue (Larréché and Moinpour, 1983, Agnew 1985, cited in Armstrong, 1989) suggest that the median would improve accuracy. Certainly, there are situations where one method is more accurate than another. (Armstrong, 1989)

**Value of Expertise in Judgmental Forecasts**

An interesting issue is how much expertise is needed for judgmental forecasting. Surprisingly, research to date (Armstrong, 1985, pp. 91-96) indicates that high expertise in the subject area is not important for judgmental forecasts of change. It is, however, important for assessing current levels. An important conclusion, then, is not to spend heavily to obtain the best experts in the field to forecast change. But one should avoid people who clearly have no expertise.

Extensive research over the last two decades has examined biases that occur in judgmental forecasting. Among these biases are optimism, conservatism, anchoring, and an overemphasis on easily available data. While some sources of bias have been identified, little knowledge exists as to how these biases affect marketing forecasts. (Armstrong, Brodie & McIntyre, 1987)

When using experts, it is essential to bear in mind that people who hold viewpoints on an issue tend to perceive the world so as to reinforce what they already believe; they look for "confirming" evidence and avoid "disconfirming" evidence. There is much literature on this phenomenon, commonly known as "selective perception." (Armstrong, 1978) In cases where disconfirming evidence is thrust upon people, they tend to remember incorrectly. Fischhoff and Beyth (1975), for example, found that subjects tended to remember their predictions differently if the outcome was in conflict with their prediction. (Armstrong, 1978)

Experts are typically overconfident (Arkes, 2001). In McNee's (1992) examination of economic forecasts from 22 economists over 11 years, the actual values fell outside the range of their prediction intervals about 43% of the time. This occurs even when subjects are warned in advance against overconfidence.

Fortunately, there are procedures to improve forecasts by experts. A commonly used technique is to ask experts to write all the reasons why their forecasts might be wrong (Arkes, 2001). Alternatively, use the devil's advocate procedure, where someone is assigned for a short time to raise arguments about why the forecast might be wrong. However, playing devil's advocate does make the person unpopular with the group. Still another way to assess uncertainty is to examine the agreement among judgmental forecasts. For example, Ashton (1985), in a study of forecasts of annual advertising sales for Time magazine, found that the agreement among the individual judgmental forecasts was a good proxy for uncertainty. (Green & Armstrong, 2005)

If we take Bayes's theorem as the standard, people tend to adjust their predictions less than they should when they receive new information (Edwards 1982). When they consider the likelihood of an outcome from a multistage process (Hitler invades Belgium, he succeeds, Britain declares war, Hitler attacks Britain) people have the opposite tendency: they act as though their best guesses of what will happen at early stages are certainties (Gettys, Kelly, and Peterson 1982).

Stewart (2001) found that judgmental forecasts are likely to be unreliable when (1) the task is complex, (2) there is uncertainty about the environment,

(3) information acquisition is subjective, or (4) information processing is subjective. Stewart's four conditions for unreliability are likely to be met with the forecasting framework for specialty products. (Green & Armstrong, 2004)

People are willing to pay heavily for expert advice. However, expertise beyond a minimal level is of little value in forecasting. This conclusion is both surprising and useful, and its implication is clear: Don't hire the best expert, hire the cheapest expert. (Armstrong, 1980)

"Expertise ... breeds an inability to accept new views." - Laski (1930)

Although experts are poor at forecasting, this does not mean that judgmental forecasting is useless. However, since all available evidence suggests that expertise beyond an easily achieved minimum is of little value in forecasting change, the most obvious advice is to hire inexpensive experts. Also, look for unbiased experts – those who are not actually involved in the situation. Finally, there is safety in numbers. Robin Hogarth (1978) has suggested using at least three independent experts and preferably six to ten!

Some expertise seems to lead to a higher level of accuracy in forecasting change; beyond a minimal level, however, additional expertise does not improve accuracy – and there is even some evidence that it may decrease accuracy. (Armstrong, 1980)

**Evaluating the utility of the Forecasting Model**

The usefulness of a quantitative model depends on both "acceptability" and "quality." Acceptability refers to approval by those who would actually use the model, while quality refers to the ability to provide better predictions or decisions. A model must score well on both characteristics if it is to be judged useful. A high-quality model that is not accepted is of no value. Usually, some trade-offs must be made between quality and acceptability. (Shapiro and Armstrong, 1974)

A model is said to be "good" if it is better than alternative models. Quality and acceptability are characteristics that may depend not only upon the model but also upon the situation. (Shapiro and Armstrong, 1974)

Research in forecasting has commonly assumed that accuracy is the primary criterion in selecting among forecasting techniques. In fact, it has been used as the sole criterion in many studies. In the sixteen 1992 International Journal of Forecasting papers that compared the results of different techniques and series, only one used criteria other than accuracy.

When asked 'Relative to other considerations (e.g. cost, ease of interpretation, cost/time, ease of use), how important is the accuracy of the forecasting methods that you use?' 29% of the experts said that accuracy was 'extremely important' and an additional 56% said that it was 'important'. These results are similar to the opinions of practitioners and researchers as reported in Carbone and Armstrong (1982) and with those of practitioners as reported by Mentzer and Cox (1984).

However, this single-minded focus on accuracy is not completely reasonable.

To encourage diffusion, new techniques should be evaluated, not only in terms of comparative accuracy, but also in terms of the "ease of use," "ease of interpretation," and "flexibility." "Cost savings" varied in rank depending upon its framing from a top criterion if related to savings from improved decisions to a lower criterion if linked to savings from technique development and maintenance. (Yokum and Armstrong, 1995)

Witt and Witt (1992) found that "speed" was most important for short-range forecasts, while "accuracy" was most important for medium- and long-term forecasts.

The evaluation of overall quality of the model calls for an examination of four key stages.

The first stage relates the "real world" to the assumptions of the model: Are the assumptions reasonable and comprehensive? A review of written documents must be carried out in order to develop an explicit listing of the key assumptions. This list may be checked by conducting interviews with the advocates of the model. The assumptions are then tested for reasonableness against: (1) empirical evidence, (2) judgments of managers, and (3) assessments by the evaluator. Admittedly, this procedure is rather crude; however, the objective at this stage is merely to identify "highly unreasonable" assumptions. (Shapiro and Armstrong, 1974) Their appeal was strictly one of face validity—that is, the assumptions seem reasonable. (Shapiro and Armstrong, 1974)

The second stage relates the model's assumptions to the final form of the model. Does the model follow logically from the assumptions? This is an examination of the logical structure of the model. (Shapiro and Armstrong, 1974) This stage of analysis is generally the most important one for assessing the quality of a model. One possible approach is to assess the total costs associated with the model [Initial development (money and time), Maintenance (money and time), User (ease in understanding, time to get results, need for expert assistance)] versus the total benefits derived [Predictive accuracy, Ability to assess uncertainty, Identification of improved policies, Learning (the model improves as experience is gained), Ability to assess effects of alternative policies, Adaptability (can adapt as the environment changes)]

The third stage relates the model and its outputs: Given the same input data, can the outputs be replicated? And the fourth stage relates the outputs to the real world: Do the benefits of the model (e.g., better predictions, better assessments of risk, or better decision making) justify the costs of the model? (Shapiro and Armstrong, 1974)

Based on the foregoing sections that review empirical forecasting literature, a **Summary of General Principles** to be used while developing forecasting procedures (Green & Armstrong, 2005) is summarized below:

(i) Domain knowledge should be incorporated into forecasting methods. In this study, before embarking on the process of constructing a demand forecasting model, a conceptual framework that charts the various variables that influence the demand in turbulent markets.

A concept map draws on studies by several experts. Moreover, when respondents are contacted for in depth interviews and focus groups during the qualitative phase of the study, domain knowledge gained will be validated with practical insights.

(ii) When making forecasts in highly uncertain situations, be conservative. Forecasts will provide for pessimistic, optimistic and average demand scenarios. This is considered essential considering the dynamic environment of markets and the complex interplay of several causal variables.

(iii) Complex methods have not proven to be more accurate than relatively simple methods. Given their added cost and the reduced understanding among users, highly complex procedures cannot be justified. Simplicity in data collection methods is considered essential to keep the consumers at ease and solicit accurate responses. Response scales will also be kept short and simple to understand.

(iv) In case data on actual behaviour is unavailable, forecasts based on judgments or intentions, may be used to predict behaviour. Past demand data for new products and unorganised markets is unavailable from reliable sources. We will therefore attempt to construct demand estimates based on a survey of buyer intentions.

(v) Methods that integrate judgmental and statistical data and procedures can improve forecast accuracy.

In the next chapter, the forecasting method suggested shall integrate intentions data gathered in terms of 'probability of purchase' and combine it with simple statistical procedures in order to estimate demand in the near term.

(vi) When making forecasts in situations with high uncertainty, use more than one method and combine the forecasts, generally using simple averages.

*This review of literature on forecasting techniques and forecasting imperatives provides the overall framework that will be used to construct, apply and test the demand forecasting model for new and turbulent markets.*

# CHAPTER 6

# Building Forecasts, Without Past Data

Forecasts typically rely on past data and empirical trends. However, in several scenarios, past data is either unavailable, inaccurate, incomplete or irrelevant. Hence, building a reliable forecast even without past data is an essential skill required by most industries, especially in the post pandemic world.

**Table 5. Data Gaps**

| Nature of data | Reason for data gaps |
|---|---|
| Unavailable data | New companies, unorganised sectors |
| Inaccurate data | Poor data collection and analysis, unreliable secondary data, lack of validation of open source information |
| Incomplete data | Inadequate sample sizes, geographical or conceptual delimitations to the study (deliberate or otherwise) |
| Irrelevant data | Post pandemic and post-recession business environment cannot use data from the immediate past periods |

**Current Intent as a Proxy for Past Data**

In the absence of usable past data, a possible alternative is usage of **current intentions as a proxy**. Current intentions capture a potential buyer's own past experiences with the product or service or an understanding of others' experiences. Before a potential buyer decides on a purchase, he would also consider his past experience with alternative products or services as well as substitutes. Hence if we can use a systematic process to decipher current intent and understand why a specific purchase decision was made, this would serve as a mind-map that captures the rationale underlying the consumer's purchase decision. This would then lead to the **identification of the core factors that influence demand for the product or service**. A comparison of these factors with the features of the product or service being offered would identify if the buyer's needs and the product features have a **good fit**, hence identifying **potential buyers** and their **probability of making a purchase**.

**A Step by Step Guide to building a Near Term Forecast, without Past Data**

The following are the key steps required to develop a near term forecasting model for new products or specialised products, without using any past data. Alternatively, this model will use current intentions as a proxy for past data.

- Define the target market clearly
- Suggest a profile of the typical potential user for the product (or service)
- Check levels of awareness about the product (or service) within the target market
- Check how many have tried or witnessed someone else try the product (or service) in the target market
- Identify the factors that influence demand for the product (or service) in the target market, and assess the importance of each feature
- Map the similarity between the influencing factors and the specific features of the product (or service)
- Assess the extent of the relationship between the causal factors and demand for the product or service. This can be done by asking the probability of purchase, should the feature exist.
- Ascertain the ideal purchase frequency of the product (or service). This can be defined by the seller based on science or general consumer habits.
- Use the above to generate a predictive model to forecast demand for the short / near term
- Calculate the error in the estimate with successive iterations

**Step 1: Define the target market clearly**

The target market is the expected 'universe' of the product's potential buyers. As a first step, this market needs to be accurately defined. A wrong definition of the market will lead to a business creating erroneous forecasts.

To illustrate, Richard Branson's Virgin Group defines its market as 'the market to get its consumers to have FUN'. Any product (or service) that fits this definition of 'enabling FUN' is fair game for the Virgin group companies. Hence, Virgin Airlines, Virgin Music, Virgin Galactic all dig into the common 'fun wallets' of their potential buyers and attempt to maximize the company's share of this wallet. Had the group defined its market differently, their entire product portfolio and promotion strategy would have radically changed.

**Step 2: Suggest a profile of the typical potential user for the product (or service)**

A product developer must create a potential buyer's profile in as much detail as possible. This should be based on whose habits were considered while designing the product.

| Product | Consumer Profile |
|---|---|
| Indigo Airlines | Budget-conscious leisure and business travelers who value affordability, efficiency, reliability, a hassle-free flying experience without compromising on quality |
| SkyDive Dubai | Adventure Seekers, Travel Enthusiasts, Adrenaline Junkies, Celebration and Gift Seekers, Bucket Listers |
| The Nothing Phone | Minimalists, individuals who value simplicity, those who appreciate a clean and uncluttered design that focuses on functionality and user experience, Design Conscious Consumers, Privacy Advocates, Environmentally Conscious |

The demographics (age, gender, income, geography, family life cycle stage, etc.) can be matched to the habits which make the product relevant and added to the intended buyer profile.

| TG | Based on secondary data available and the buyer profile thus identified, the **TARGET POPULATION GROUP** of potential buyers is calculated and forms the universe for the product (or service) offering. |
|---|---|

For the more statistically inclined, once the reasons why a product or service will be purchased is ascertained, discriminant analysis may be used to identify the characteristics of a consumer who is most likely to purchase the product or service. The discriminant analysis would be used to create a model to classify future users and non-users of the product or service.

> *Discriminant analysis* is a statistical technique used to classify observations into different groups based on a set of predictor variables. It helps us understand how different variables contribute to distinguishing between groups. Discriminant analysis can help businesses identify different groups or segments of customers based on their characteristics, such as age, income, and purchasing behaviour. This information can then be used to target specific product offerings to each segment based on the 'intent to buy', improving customer satisfaction and overall business performance.

**Step 3: Check levels of awareness about the product (or service) within the target market**

If a potential buyer is not aware about the product, there can never be any demand. Once the buyer profile is created, a set of potential buyers can be identified and a sample of them can be asked whether they are aware about the product or not. The percentage of those who are aware of the product (or service) narrows down the effective target market.

*For example, based on the profile developed in Step 2, Indigo Airlines, which is opening its operations in the temple town of Ayodhya for the first time, identifies 100 persons who fit the potential buyer profile and ask them if they are aware about Indigo. If 20% of the sample are unaware about the existence of the airline, then they can never be part of the effective target market in Ayodhya town.*

**AWARENESS** is checked off in the survey responses as a dichotomous variable with A=0 if awareness is not there and A=1 if the respondent is aware of the product/service. A percentage of the respondents aware about the product are then ascertained [A%].

> A %

### Step 4: Check how many have tried or witnessed someone else try the product (or service) in the target market

After checking Awareness levels, one's own past trial/usage of the product or observing someone else try/use the product (or service) helps build the demand estimate further. One's own usage or observation of someone's usage can be ascertained with a follow up question.

> T %

If a respondent is aware about the product, they can be next asked if they have used/seen someone else use (T=1), or not used/not seen some else use the product (T=0). A percentage of the respondents who tried/witnessed a **TRIAL** of the product is then ascertained [T%].

**Unlike traditional methods that extrapolate from the past, we begin this forecasting process with an assessment of the present level of usage of the product or service in each potential buyer segment.**

*The potential buyers pool will therefore be:*

$$ \boxed{TG} \times \boxed{A\%} \times \boxed{T\%} $$

It is noteworthy that this is a multiplicative function and hence at any stage if the Awareness or Trial levels are zero, the entire demand forecast reduces to zero.

**'Witnessing someone else try':** It is also essential to understand who decides on the purchases of similar products or services, who or what influences the purchase decision the most? Listing of all influencers and gaining an indication of the most crucial influencer in the decision making process becomes crucial.

This will help appreciate the dynamics that are involved in the decision to purchase similar products and services. Typically, gathering expert opinion using focus groups is an ideal method for validation of the factors identified and will also provide deeper qualitative insights.

The model thus factors in the fact that observing an influencer or expert or even a friend or family member also triggers demand and increases the 'T%'.

***Let's take an example to illustrate this process further:***

*For an Android application, the target potential user base in the near term would be Smartphone owners with Android Operating systems connected to an active Data Plan. [TG]*

*A sample of smartphone owners having this basic profile could be asked about their awareness about the specific app for which the demand forecast is being developed [A%]*

*Thereafter, application usage patterns and their responses would help estimate the proportion of potential buyers having used or having no usage of similar Android applications. Observing 'someone else try' may also be considered adequate. Witnessing a demo on the day of the survey may also increase this 'Trial' ratio. [T%]*

*This part of the demand forecasting process may be done using a structured questionnaire, online poll or similar structured data collection instrument.*

**Step 5: To identify the factors that influence demand for the product or service**

Factor Analysis can then be used to narrow down to the key factors that influence purchase of the product or service.

> *Factor analysis* is a statistical technique used to understand the relationships between a set of observed variables and the underlying latent factors that may be influencing them. It helps in identifying the underlying structure of a dataset by reducing the number of variables into a smaller set of factors.
>
> Imagine you have a large dataset with multiple variables, and you want to find out if there are any hidden patterns or factors that explain the relationships between these variables. Factor analysis can help in uncovering these underlying factors by grouping together variables that tend to co-vary.

A potential user must be asked to provide their ratings on a suitable point scale indicating the level of importance (No impact – Low impact – Medium impact – High impact) that each of the influencing factors has on their demand for the product or service.

If the product (or service) has a finite list of features (or if one does not have the skills to apply the Factor Analysis technique), a listing of all features can be made and respondents can be asked:

"Given the existence of this particular feature, what is the probability that you will buy this product?"

"Rank the importance of this feature in your purchase decision?"

**Back to our example to describe the process further,**

*For the smartphone application, a set of questions could include the following –*

| X (0,1) | Level of importance in influencing purchase  W | | | |
|---|---|---|---|---|
| *Features* | *No impact* | *Low impact* | *Medium impact* | *High impact* |
| | *0* | *1* | *2* | *3* |
| *The application is compatible for Android Operating system* | | | | |

| | | | | |
|---|---|---|---|---|
| *The application is available free of charge* | | | | |
| *The application is advertised widely on social media platforms* | | | | |
| *The application has a high ranking on web reviews by existing users* | | | | |
| *The application has a high ranking on web reviews by critics* | | | | |
| *The application is downloadable within 2 minutes on an average Wi-Fi network* | | | | |
| *The application has high privacy settings* | | | | |
| *The application has a user friendly interface* | | | | |
| *The application download size is less than 100 MB* | | | | |
| *The application is available in regional languages* | | | | |

**Note:** The features listed are ALL the features that are identified as 'reasons to buy' the type of product. Some may or may not exist in the specific product for which a demand forecast is being prepared. Hence 'F=0' if the feature DOES NOT EXIST in the product for which the forecast is being prepared and 'F=1' if the feature EXISTS in that product. If F=0 demand will not be generated due to that particular feature, since it does not exist in the product under consideration. If the feature exists, that will be an influencer of demand. **[X(0,1)]**[1]

The Intentions survey also asks the consumer about the **level of importance of each feature** on a scale of 0 to 3 (see table above), with 0=no impact and 3=high impact on the purchase decision. The average of these scores will provide an IMPORTANCE SCORE / WEIGHTAGE to the specific features listed. **[W]**[2]

The consumer can then be asked *'If this feature exists, what is the likelihood of them purchasing the product (or service)?'*. The responses can be on a scale of 0% to 100% and will thus create the **'probability of purchase'**, given the particular feature exists in the product. [P][3]

**Let's return back to our example to illustrate:**

| Features | [1] **Exists in product? (Yes/No) [X(0,1)]** | *Level of importance in influencing purchase (on scale of 0 to 3)* | | [3] **Average Probability of purchase, if feature exists [P]** |
|---|---|---|---|---|
| | Yes = 1 No = 0 | [2] **Average Rating [W]** | Ranking | 0% to 100% |
| *The application is compatible for Android Operating system* | 1 | **2.8** | III | **100%** |
| *The application is available free of charge* | 1 | **2.4** | IV | **90%** |
| *The application is advertised widely on social media platforms* | 1 | **1.5** | VIII | **62%** |
| *The application has a high ranking on web reviews by existing users* | 1 | **2.2** | V | **78%** |

---

[1]   *The existence of the respective features is based on the manufacturer's catalogue/ declarations*

[2]   *Average Rating considers the simple mean of the ratings assigned to each factor by all respondents*

[3]   *Average Probability is the simple mean of the likelihood of purchase, subject to the respective feature being present in the product, as opined by all respondents*

| The application has a high ranking on web reviews by critics | 0 | 2.7 | -- | 65% |
|---|---|---|---|---|
| The application is downloadable within 2 minutes on an average Wi-Fi network | 1 | 2.0 | VI | 91% |
| The application has high privacy settings | 1 | 2.9 | II | 88% |
| The application has a user friendly interface | 1 | 3.0 | I | 100% |
| The application download size is less than 100 MB | 1 | 1.8 | VII | 82% |
| The application is available in regional languages | 0 | 2.5 | -- | 55% |

**Using the above intentions survey data, gathered by seeking responses from a sample of potential buyers, the weighted average probability of purchase is calculated in this step as follows:**

$$\text{Weighted average probability of purchase} = \left( \frac{\sum_{i=1}^{k} W_i X_i P_i}{\sum_{i=1}^{k} W_i} \right)$$

| Features | [1]Exists in product? (Yes/No) [X(0,1)] | Level of importance in influencing purchase (on scale of 0 to 3) | [3]Average Probability of purchase, if feature exists [P] | $W_i X_i P_i$ |
|---|---|---|---|---|
| | Yes = 1 No = 0 | [2]Average Rating [W] | 0% to 100% | Columns 1 X 2 X 3 |
| The application is compatible for Android Operating system | 1 | 2.8 | 100% | 2.8 |
| The application is available free of charge | 1 | 2.4 | 90% | 2.2 |
| The application is advertised widely on social media platforms | 1 | 1.5 | 62% | 0.9 |
| The application has a high ranking on web reviews by existing users | 1 | 2.2 | 78% | 1.7 |
| The application has a high ranking on web reviews by critics | 0 | 2.7 | 65% | -- |
| The application is downloadable within 2 minutes on an average Wi-Fi network | 1 | 2.0 | 91% | 1.8 |

| | | | |
|---|---|---|---|
| The application has high privacy settings | 1 | 2.9 | 88% |
| | | | 2.55 |
| The application has a user friendly interface | 1 | 3.0 | 100% |
| | | | 3.0 |
| The application download size is less than 100 MB | 1 | 1.8 | 82% |
| | | | 1.5 |
| The application is available in regional languages | 0 | 2.5 | 55% |
| | | | -- |
| 'k' = number of features (Here, k = 10) | | $\sum\limits_{i=1}^{k} W_i$ = 23.8 | | $\sum\limits_{i=1}^{k} W_i X_i P_i$ = 16.45 |

$$\text{Weighted average probability of purchase} = \left( \frac{\sum\limits_{i=1}^{k} W_i X_i P_i}{\sum\limits_{i=1}^{k} W_i} \right) \begin{array}{l} = 16.45 \div 23.8 \\ = 69.1\% \end{array}$$

Thus, we have now assessed the levels of influence that each of the features (or 'influencing factors') has on the demand for the product (or service).

A Survey of Buyer Intentions has now assessed the **'weighted average probability of purchase'** using end user's probability ratings assessing the probability of purchase of the product or service, given the existence of each factor identified as crucial to the buying decision.

The intentions survey has gathered this information by –

    (i)   Asking potential buyers about the likelihood of their purchase of the product or service, subject to the identified feature being present.

    (ii)  Evaluating the importance of each feature based on the potential buyer's preference

    (iii) Combining the above information to generate a 'weighted average probability of purchase' estimate

**Step 6: To use all of the above to generate a near term predictive model to forecast demand for the product or service**

Based on the study, a forecasting equation may be prepared for the target group of potential buyers to estimate near-term demand for the product (dependent variable), given existence and impact of each causal variable (features).

Pessimistic, optimistic and average estimates of demand for the product or service may also be generated, in quantity and market value terms.

$$\text{Demand forecast (units)} = TG \times A\% \times T\% \times \left( \frac{\sum_{i=1}^{k} W_i X_i P_i}{\sum_{i=1}^{k} W_i} \right)$$

*(Refer Steps 2, 3, 4 and 5 above)*

*If the Target Group of potential buyers for the Android application (in our example) is 10,000 people, of whom 80% are aware of the product, and 70% of those who are aware have tried (or witnessed someone try) the product, then after assigning the weighted average probability of purchase (69.1%) based on the calculation above, the Demand forecast, will be:*

**Demand forecast** (units) = 10000 X 80% X 70% X 69.1% = 3870 users

Assume that the cost of the Android app is Rs 100 per user per month, the forecast sales revenue can be calculated as below:

**REVENUE FORECAST = Demand Forecast (Units) X Usage frequency per year X Cost per use**

$$= 3870 \text{ users X } 12 \text{ months X } 100 \text{ Rs/month}$$
$$= \text{Rs } 46,44,000$$

**Step 7: To estimate the accuracy of the demand estimate**

The overall accuracy of the estimate of demand for the product or service may be ascertained by comparing demand estimates from other sources and evaluate model accuracy. Alternatively, an error term may be estimated by reviewing how many potential buyers who showed high probability of purchase in the forecasting period actually made a purchase by the end of the period. Over multiple iterations, this error term that adjusts for overstated and understated intentions will stabilize. However, no such error estimate would be available for the first forecasting period.

While assessing intent, it may be possible that potential buyers **over-state or under-state their intentions**. A comparison between the stated intent and actual purchase behaviour in a defined period would need to be tracked using the company's sales trackers and management information systems. The difference between the potential forecast and actual demand will lead to an error term. This will stabilize over multiple iterations for similar sectors and can be used as an error estimate in the forecast. [± e]

$$\textbf{Demand forecast (units)} = \text{TG} \times A\% \times T\% \times \left( \frac{\sum_{i=1}^{k} W_i X_i P_i}{\sum_{i=1}^{k} W_i} \right) \pm e$$
*(Adjusted)*

*As per our example*, 69.1% is the weighted average probability of purchase as estimated by the intentions survey. There will be no error estimate in the first instance of using this forecast.

However, when the 12 month period ends, a set of or all the respondents selected from the sample set used during the forecast building process can be re-approached and asked if they actually did make the purchase. Alternatively, the internal records of the company may also be used to check actual purchase patterns during the forecast period. If only 60% of those surveyed end up actually buying (versus the 69.1% who said they would), then this is a case of over-stated intentions and the error rate is **9.1%**

Similar error rates can be checked over multiple years, and it will most likely be observed that the error term stabilizes over multiple iterations. If the error rates in the following two more forecast years end up to be 2.5% and 6.2%, then the average Error estimate '±e' will be **±5.9%**

**Demand forecast range** (units) = 3870 users **±5.9%**

$$= \textbf{3642 users (minimum) to 4098 users (maximum)}$$

Assume that the cost of the Android app is Rs 100 per user per month, the forecast sales revenue can be calculated as below:

**REVENUE FORECAST = Demand Forecast (Units) X Usage frequency per year X Cost per use**

Minimum revenue forecast = 3642 users X 12 months X 100 Rs/month = Rs 43,70,400

Maximum revenue forecast = 4098 users X 12 months X 100 Rs/month = Rs 49,17,600

Thus, factoring the error term, the company can set a sales forecast, based on the above demand estimates, ranging between **Rs. 43.70 lakhs to Rs 49.17 lakhs.**

**Limitations of this Forecasting Model**

**Relevance**

This model can be applied to industries or for any product where building awareness about the concept and its utility is critical. It cannot be applied to impulse buys or very low value purchases where the consumer does not pause to specifically evaluate influencing factors.

The assessment of purchase intentions would be relevant for a consumer group within the target area of the study. There may be certain changes in these estimates and weightages within other areas and across other geographies. Considering this limitation, the model has been applied and demand estimates prepared only for the area where the purchase intentions have been gathered from.

The assessment of difference between actual purchase behaviour and stated intentions is critical to the accuracy of the model. This 'adjustment factor' addresses issues of affordability, availability, ability and willingness to buy as well as under and overstated intentions. In case this factor is not available to users of the model, the accuracy of the demand estimate would be lower.

The qualitative phase of research may identify a few factors that were not specifically included in the survey questionnaire. 'Probability of purchase' estimates exclude the impact of these variables. The factor analysis results must however indicate that at least two thirds of the influencing variables have been explained by the survey. The impact of the remaining variables could perhaps be identified in a follow up study and included as additional variables within the forecasting equation.

It must be however borne in mind that while adding and assessing the impact of more variables would increase accuracy of the model, it would add to respondent fatigue and model complexity.

**Forecasting Period**

**This model is useful to estimate demand in the short term only.** It is dependent on a survey of buyer intentions which gets more subjective as the time horizon increases. A respondent cannot be expected to give an indication of the probability of purchase of a product beyond the near term (say six months to one year). Future research could seek to extend this model to try and develop demand estimates for the medium and long term.

**Test of the Model**

One way to assess accuracy of a forecasting model has been to examine the agreement among judgmental forecasts. For example, Ashton (1985), found that the agreement among the individual judgmental forecasts was a useful proxy for accuracy.

**Changing factors influencing demand**

Factors that affect the demand for the product and their levels of importance may differ across countries. They will also change over time. Some factors may lose their importance. New factors may emerge. Continuous research on the identifying the nature, evaluating the impact and estimating the importance of these factors that affect demand would need to be conducted to keep the forecasts accurate.

Moreover, the model can also be applied to any other industry where the product or service being offered is new and concept marketing is involved. Application of the model across industries with a portfolio of specialty offerings will be useful.

Another area of continuing importance would be to continuously track the intentions survey data and actual purchase patterns. Over time, it may become possible to estimate (within confidence limits) this level of difference between stated intentions and actual buying behaviour. Recognizing that the profile of buyers is different across different industries, it is only fair to suggest that these estimates would be largely industry or sector specific. However, attempts to arrive at such an estimate and also identify the factors that cause this difference would be very useful.

**The general requirements for applicability of this forecasting model are that:**

i.   The product or service is not a commodity or an impulse purchase item.

ii.  It must have a specific, identifiable usage and benefit which can be demonstrated by means of a trial or demonstration.

iii. The product should be considered 'important' by the consumers, either due to the extent of monetary outlay required to purchase the product or due to its utility.

iv.  The markets should typically be in the introductory and growth stages. As markets become mature, concepts tend to get 'commoditized', usage of the product becomes part of the consumers' habits, routinized buying behaviour sets in and differences in the relative importance of factors blur. As an assessment of the impact of the influencing factors on the purchase decision is required, the model cannot be applied when markets reach maturity. This is because, when markets reach this stage, consumers purchase the product by force of habit and do not pause to think of 'influencing factors'.

*NOTE:*

*A WEB APPLICATION FOR READERS TO READILY APPLY THIS STEP BY STEP METHODOLOGY IS AVAILABLE. Login to EmpowerToEnslave.com and register yourself to use this Forecasting Algorithm.*

# CHAPTER 7

# Enslaving With Pride

We began with the notion that markets are now in a state of perpetual motion. Products and services that are designed after understanding habits of the target consumers are most likely to see the most success. Customization most certainly leads to users feeling empowered. Empowerment builds pride and this Pride is almost always addictive. We tend to never let go of brands we love and brands that have specifically reached out to us and satisfied our ever evolving needs.

*A brand that listens and learns intently builds an 'enslaved', fiercely loyal tribe of consumers.*

**Reinforcing Consumer Empowerment**

Consumers are looking for solutions to their needs with products that can plug into or align with their habits. The following are examples of how products and services can reinforce their commitment to give the consumers what they want and thus, feel empowered.

**Customization**

Customization can take many forms across various domains, catering to individual preferences and needs. Consider how these examples from various industries have created empowered consumers, who keep coming back to the same brand.

**Gaming:** Customizable character appearances, skins, and in-game settings in video games.

**Tailoring:** Custom-made clothing tailored to individual measurements and style preferences.

**Design Your Own:** Online platforms that allow users to design their own shoes, clothing, or accessories.

**Furniture:** Customizable furniture pieces, allowing customers to choose fabrics, colors, and designs.

**Wallpapers and Paint:** Personalized wall coverings and paint colors for interior spaces.

**Car Configurators:** Online tools that enable customers to customize their cars, choosing features, colors, and accessories.

**Vehicle Wraps:** Personalized vehicle wraps and graphics.

**Beverages:** Customizable coffee blends, where customers can choose the type of beans, roast, and flavours.

**Fitness Plans:** Customized workout plans tailored to individual fitness goals.

**Nutrition Plans:** Personalized diet plans based on dietary preferences, allergies, or health conditions.

**Online Courses:** Customizable learning paths and elective courses in online education platforms.

**Music Playlists:** Customizable playlists on streaming platforms.

**Streaming Services:** Tailored content recommendations based on user preferences.

## Bespoke and Curated Collections

Bespoke and curated collections are often associated with unique, personalized, or carefully selected items that cater to specific tastes or preferences. This build cults of loyal customers who feel like a part of a unique family of users whose needs and habits, down to the minutest detail are valued by the manufacturer or service providers.

**Customized Jewellery Collection:** A bespoke collection of jewellery where each piece is individually crafted based on the customer's preferences, incorporating specific gemstones, metals, and designs.

**Tailored Suit Collection:** A bespoke collection of tailored suits, where each suit is made to measure for the client, taking into account their unique body measurements, fabric preferences, and styling choices.

**Personalized Stationery Set:** A bespoke collection of stationery items, such as notebooks, pens, and notepads, featuring personalized monograms, colors, and paper types chosen by the customer.

**Artisanal Coffee Collection:** A curated collection of specialty coffees from different regions, carefully selected for their unique flavors and characteristics, presented in a thoughtfully designed package.

**Book Club Curated Reading List:** A curated collection of books selected by a book club, featuring a diverse range of genres, authors, and themes to provide an enriching and varied reading experience for its members.

**Gourmet Food Basket:** A curated collection of gourmet foods, including artisanal cheeses, chocolates, and specialty snacks, chosen to create a luxurious and indulgent culinary experience.

**Traveler's Adventure Kit:** A curated collection of travel essentials, such as compact gadgets, versatile clothing, and innovative accessories, tailored for adventurous travelers seeking convenience and functionality.

**Create an Aura Around the Brand**

Building an aura around a brand involves creating a distinct and compelling identity that resonates with your target audience. It's about cultivating a unique atmosphere or perception that sets your brand apart and leaves a lasting impression. Any consumer who feels they are part of this identity or that the brand's identity builds their own personality status feels tremendously empowered.

To achieve this, a brand must clearly state its personality and values. It must consider what the brand stands for, its mission, and the emotions it wants to evoke. "Tajness" refers to a set of service standards and cultural values associated with the Taj Hotels, a luxury hotel chain based in India. Taj Hotels is known for its commitment to providing exceptional hospitality experiences.

A narrative that tells the compelling story of the brand helps create an emotional connection with the target audience. This storytelling may be delivered using a unique tone of voice in all communication. Whether it's formal, friendly, humorous, or inspirational, a distinctive voice helps your brand stand out and be remembered. Creating Experiences around the brand that engage the senses and emotions of your audience also leave a lasting impression.

Customers can also be part of co-creating a community around the brand. Encouraging user-generated content, creating forums for discussions, and actively engaging with the audience on social media fosters a strong sense of belonging and loyalty.

**Limited Editions and Exclusivity**

Companies introduce limited edition products or exclusive offers to create a sense of scarcity and exclusivity. This enhances the perceived value of the brand and empowers those who are able to get their hands on the exclusive offering.

LEGO occasionally releases limited edition sets, such as those tied to popular franchises like Star Wars or collaborations with iconic artists. These sets are highly sought after by LEGO enthusiasts.

Rolex, a luxury watch brand, has released limited edition watches with unique designs and features. These limited runs make the watches highly collectible and they can appreciate in value over time.

## Making Customers into Celebrities, Influencer marketing

There is no better empowerment than to be a consumer who is positioned as the face of the brand.

Instagram itself can be seen as a brand that turns users into celebrities. The platform is designed for users to showcase their lives, talents, and creativity. Users gain followers and recognition based on their content, effectively turning them into social media celebrities.

Airbnb's "Live There" campaign focuses on the idea that travelers can experience a destination like a local. By sharing personal stories and photos, Airbnb users become ambassadors of the brand, showcasing unique travel experiences.

Aries Agro Limited, which is India's largest manufacturer of specialty fertilizers, has made farmers the face of the brand during customer conventions, policy advocacy meetings with government officials and innovation conclaves, up to and including having three farmers meet President Obama during his India visit, and represent the company even without the top management present.

## Tailor-made Offers and Schemes using Retail Histories and Data Mining

Consumers may be enticed into buying with special schemes and product offers. If these are tailor-made, using data mining, user histories, location specific push notifications, then the curated offers are more valued than general advertisements and open promotions.

An E-commerce website may offer personalized discounts or promotions based on a customer's previous purchase history. For example, if a customer frequently buys athletic wear, the website might offer a special discount on new arrivals in that category.

Similarly, a hotel may provide personalized packages to guests, incorporating their preferences. This could include room upgrades, spa treatments, or dining options tailored to the guest's known interests, such as golf or cultural activities.

Mobile service providers may offer personalized data or call packages based on a customer's usage patterns. For instance, heavy data users might receive promotions for upgraded data plans.

The key to successful tailor-made offers lies in data analysis and understanding customer behavior. By leveraging customer data, businesses can create personalized experiences that not only meet individual preferences but also strengthen the overall relationship between the brand and the customer. Additionally, obtaining customer consent and ensuring data privacy are crucial aspects of implementing effective tailor-made offers.

**Engaging with Orphaned customers**

A competitor who has shut down, run out of inventory or facing supply chain issues leaves a set of consumers 'orphaned' and searching for immediate alternatives to satisfy their needs.

Seeking out such 'orphaned customers' makes them feel valued by the competition, though they were not their original clients. Opportunistically building such goodwill builds a new pool of empowered customers.

When Jet Airways and GoAir suddenly shut operations in India, there was a massive pool of passengers who were left in the lurch. Providing alternatives to these flyers after the cancellations could have built a new pool of customers. However, in contrast, the suddenly shrinking number of seats skyrocketed airfares and the remaining airlines lost the opportunity to build goodwill.

When a smartphone brand exits the market, competitors such as Xiaomi, Samsung, and Vivo may attract new customers. These brands often offer a range of smartphones at various price points, appealing to diverse consumer segments.

It's important to note that the success of capturing competitor customers depends on various factors, including the brand's reputation, product quality, pricing strategy, marketing efforts, and customer service. Additionally, new entrants or emerging brands with innovative offerings may also seize the opportunity to attract customers in the absence of a competitor.

**Geo-fencing**

Remember the time when you enter a mall and receive a WhatsApp or text message informing you of a sale taking place in a store within the premises. You feel well informed and the surprising message may lead you to visit the store. This is Geo-fencing, working in collaboration with your mobile service provider and the Mall administration.

Geo-fencing is a location-based technology that allows businesses and mobile applications to set up virtual boundaries, known as geo-fences, around specific geographic areas. These virtual perimeters can trigger actions when a mobile device enters or exits the defined area. Geo-fencing relies on GPS, Wi-Fi, cellular data, or RFID (Radio-Frequency Identification) to determine the device's location.

Providing timely information using push notifications builds well informed, empowered users.

**A-la-carte pricing**

During the product design process, features are the most important cost determinants. Ask consumers about features that they rarely or never use and eliminate them from the product design. This reduces a product's bill of materials, without compromising quality.

A base version of the product is provided and any add-on features are available at an added cost to any user who chooses to add them. This 'a-la-carte pricing' provides variants at multiple price points and consumers feel that they were asked to pay only for features that added value to them.

A-la-carte pricing may, at times, lead to realisations that are higher than the all-in-one-bundled product. A-la-carte pricing is employed in various industries to provide flexibility and customization for customers.

Cable and satellite TV providers often offer a-la-carte pricing for additional channels or premium content. Customers can choose specific channels or add-ons, giving them control over their entertainment package.

Software as a Service (SaaS) providers may offer a-la-carte pricing for additional features or advanced functionalities beyond the basic subscription. This allows customers to customize their software package based on their specific needs.

Airlines are the most common user of a-la-carte pricing for services such as seat selection, extra baggage, in-flight meals, or Wi-Fi. Travelers can choose and pay for the specific services they value, rather than being charged for a bundled set of amenities.

**These empowered consumers tend to keep coming back for more. 'They listened, they cared, they gave me exactly what I needed'.**

## Creating 'Stickiness': Enslaving Consumers

Empowerment builds a satisfied set of consumers. However, building unshakeable brand loyalty is an art form that companies must learn and use wisely. Current day marketing does not end at enlisting potential buyers and engaging them with well-crafted communication and cost effective pricing. After the sale, keeping consumers engaged and always inquisitive ensures an enviable mindshare and top-of-the-mind recall that will defend and grow market share.

## Planned Obsolescence

There is nothing more dull and dreary than a brand that has never 'changed its stripes'. Planned obsolescence is a business strategy where a product is intentionally designed to have a limited lifespan or become obsolete after a certain period. The goal is to encourage consumers to replace the product more frequently, thus driving additional sales for the manufacturer.

Products are designed to become irrelevant after a specific, planned duration of time. In this period, customers or processes have become so habituated to the product that they agree to or are left with no choice but to upgrade.

**Functional Obsolescence:** Products are designed with components that have a limited lifespan or are not easily replaceable. This ensures that the product will eventually become non-functional, prompting consumers to purchase a new one.

**Technological Obsolescence:** Products are intentionally designed to become outdated quickly due to rapid advancements in technology. This encourages consumers to upgrade to the latest models with improved features and capabilities.

**Style Obsolescence:** Products are designed to go out of style or look outdated after a certain period. This can be prevalent in industries like fashion or home decor, where trends change quickly, compelling consumers to purchase the latest designs.

**Software Obsolescence:** Software and digital products may be designed to become incompatible with older versions or devices, requiring users to upgrade to newer versions to access the latest features or remain compatible with other systems.

**Limited Repairability:** Products are designed with components that are difficult to repair or replace. This discourages consumers from attempting to fix the product and often leads them to opt for a new purchase instead.

**Incompatibility:** Products are intentionally made incompatible with newer accessories, peripherals, or technologies. This forces consumers to upgrade their entire system or purchase new accessories, contributing to the cycle of obsolescence.

**Shortened Product Life Cycles:** Companies intentionally shorten the product life cycle by introducing new models or versions more frequently. This practice is common in industries like smartphones and electronics.

### Loyalty Programmes to Reward relationships

Loyalty programs are designed to encourage repeat business and customer retention by rewarding customers for their continued engagement with a brand. These programs typically offer various incentives, discounts, or exclusive perks to loyal customers, thus 'enslaving' them.

Customers earn points or rewards for each purchase, and these can be redeemed for discounts, free products, or other benefits. Loyalty programs often have tiered structures where customers can unlock additional benefits or rewards as they reach higher levels of spending or engagement.

Airline Frequent Flyer programme, Shopping Mall reward programmes, Credit card reward points are the most common forms of loyalty programmes. Many also provide cross selling opportunities to the companies and millions of reward options for their loyal users.

Loyalty programmes involve a marketing cost but until the points are redeemed, are also a source of interest free working capital for the issuing companies. The value of the consumer and usage data is also tremendous.

### Subscription models

Subscription models involve charging customers a recurring fee at regular intervals for access to a product or service. This approach has gained popularity across various industries, offering businesses a steady revenue stream and customers the convenience of regular, automatic access to products or services.

The earliest implementers of subscription models (way before even the term was coined) in India are the daily milk, newspaper and bread delivery vendors. They sign us up for a long period of usage and ensure uninterrupted supply of these items, at a monthly fee.

Netflix users pay a monthly fee for access to a library of digital content, such as movies, TV shows, and documentaries.

Xbox Game Pass gamers pay a subscription fee for access to a library of video games, which they can play as long as their subscription is active.

Food Darzee offers home delivery of well-balanced healthy meals for assisting the fitness and wellness goals of its clients in major cities of India.

Such subscriptions keep consumers captive, until they choose to opt out of the service. However, there is nothing more habit forming than a prepaid subscription, with an auto debit setup on your credit card for renewals!

**Incentivize repurchase**

Encouraging customers to make repeat purchases not only increases revenue but also builds long-term relationships.

Brands must stay connected with customers after the sale through follow-up emails, newsletters, or targeted communication. Sharing relevant content, product recommendations, and exclusive offers keep users engaged and tuned in.

Gamification can add a new dimension into the shopping experience, such as earning badges, levelling up, or participating in challenges. This makes the repurchase process more enjoyable and rewarding.

## Upselling

Upselling is a strategy where a business encourages customers to purchase a higher-end or more expensive product or service than the one they initially intended to buy. The goal is to increase the average transaction value and maximize revenue from each customer.

**Bundle Offers:** Combining related products or services into bundles and offering them at a discounted price when purchased together not only increases the overall value but also provides customers with a perceived deal.

**Upgrade Options:** Present customers with upgraded versions of the product or service they are considering, highlighting additional features, enhanced performance, or improved specifications.

**Cross-Selling:** Recommend complementary products or services that go well with the customer's original purchase. For example, if someone is buying a camera, suggest additional lenses, accessories, or a camera bag.

**Customization Options:** Provide customization options for products or services, allowing customers to tailor their purchase to better suit their needs. Offer premium customization features at an additional cost.

**Risk-Free Trials:** Offer a risk-free trial or money-back guarantee for the upgraded product or service. This reduces the perceived risk for customers, making them more likely to opt for the higher-end option.

**Highlight Cost Savings:** Demonstrate how investing in a higher-priced product can lead to long-term cost savings. For example, a more expensive appliance may be energy-efficient and result in lower utility bills over time.

**Personalized Recommendations:** Use customer data to provide personalized recommendations based on their preferences, purchase history, or browsing behavior. Tailor the upsell to meet their individual needs and interests.

**Add-Ons at Checkout:** During the checkout process, suggest relevant add-ons or accessories that enhance the customer's experience with the main product. Ensure that the suggested items are closely related to the customer's initial purchase.

**In-Store Demonstrations:** In a retail setting, conduct in-store demonstrations or product trials that showcase the benefits of higher-priced items. Allow customers to experience the added value before making a decision.

**Upgradable Features:** Design products with upgradable features or modular components. Customers can start with a basic version and later upgrade specific elements without having to replace the entire product.

Effective upselling however requires that the customer sees added value, rather than simply increasing the price. By understanding customers' needs and preferences, businesses can offer upsells that genuinely enhance the customer experience and satisfaction.

Striking a fit between what your customer wants, what your brand delivers and knowing what your competitor's brand can never provide will help companies identify their **'win zones'**

Customer Enslavement is real and possible. However, few include Empowerment and Enslavement specifically into their marketing strategy designs. 'Enslavement' is a shock-and-awe way of expressing 'fierce customer loyalty' towards a brand. It is indeed, the best way to ensure business continuity and growth.

# REFERENCES

**Agricultural Economics**

Allen, P. G. (1994). "Economic Forecasting in Agriculture". *International Journal of Forecasting,* 10 (1994) pp. 81-135.

Armstrong, Scott J. (1994). "The Fertile Field of Meta-Analysis: Cumulative Progress in Agricultural Forecasting." *International Journal of Forecasting,* 10 (1994), pp. 140-147.

Singh, Rakesh (2005). "Predicting demand in an uncertain world." Hindu Business Line www.thehindubusinessline.com/catalyst/2005/10/06/stories/2005100600060200.htm


**Consumer Behaviour and Marketing**

Agarwal Nitin & Agarwal Manish (2003). "Theory of Trying – Implications for Marketing New-Concept Products." *IIMB Management Review,* Dec 2003, pp. 15–21.

Bagozzi, Richard P, and Paul R Warshaw, (1990). "Trying to Consume", *Journal of Consumer Research,* 17, Sept, pp 127-133.

Bentler, P. M. & G. Spechart (1979). "Models of Attitude-Behavior Relationships," *Psychological Review,* 86, pp 452-464.

Bomberger, William A. (1996). "Disagreement as a Measure of Uncertainty." *Journal of Money, Credit and Banking* 28 (Aug 1996): pp. 381-392.

Carroll, Christopher D. (2002). "The Epidemiology of Macroeconomic Expectations". Center on Social and Economic Dynamics, CSED Working Paper No. 25. http://www.brookings.edu/es/ dynamics/papers/epidemiology/epidemiology.htm

Fishbein, M. & Ajzen, I. (1975). *Belief, Attitude, Intention, and Behavior.* Reading, MA: Addison-Wesley.

Hair, J.F., Bush, R.P., Ortinau, D.J. (2006). *Marketing Research within a Changing Information Environment*, Third edition. Tata Mc-Graw Hill Publishing Company Ltd., New Delhi.

Hawkins, D.I., Best, R.J., Coney, K.A. (2002). *Consumer Behaviour: Building Marketing Strategy.* Eighth Edition. Tata McGraw-Hill Publishing Company, New Delhi.

Larréché, Jean-Claude and R. Moinpour (1983). "Managerial judgment in marketing: The concept of expertise," *Journal of Marketing Research*, 20,110-121.

Malhotra, Naresh K. (2001). *Marketing Research: An applied orientation.* Third edition. Pearson Education.

Morrison, D. G. (1979). "Purchase intentions and purchase behavior," *Journal of Marketing*, 43, 65-74.

Shapiro, Alan C. and Armstrong J. Scott (1974). "Analyzing Quantitative Models." *Journal of Marketing*, Vol. 38, No. 2, (1974), pp. 61-66

Silk, S. J. & G. L. Urban (1978). "Pre-test market evaluation of new packaged goods: A model and measurement methodology," *Journal of Marketing Research*, 171-191.

Solomon, M.R. (2004). *Consumer Behaviour: Buying, Having and Being.* Sixth edition. Pearson Education.

**General Management & Strategy**

Armstrong, J. Scott (1967). "Derivation of Theory by Factor Analysis." *The American Statistician*, 21 (December), pp 17-21.

Ashton, Alison H. (1985). Does consensus imply accuracy in accounting studies of decision making? *Accounting Review* 60, 173-185.

Callopy, Fred and Armstrong J. Scott (1992). "Expert Opinions About Extrapolation and The Mystery of the Overlooked Discontinuities." *International Journal of Forecasting* 8, pp 575-582

Edwards, W. (1982). "Conservatism in human information processing", in Kahneman, D., Slovic, P. & Tversky, A. (eds.), *Judgment under uncertainty: heuristics and biases*. Cambridge, UK: Cambridge University Press.

Emery, F. E. and Trist, E. L. (1965). "The Causal Texture of Organizational Environments," *Human Relations*, 18(1), pp 21-32.

Hogarth, Robin M. (1978). "A note on aggregating opinions," *Organizational Behaviour and Human Performance,* 21: pp 121-129.

Khandwalla, P. N. (1977). *The Design of Organization*, New York: HBJ.

McLarney Carolan (2003). "Strategic Planning Processes in Chaotic Environments: How to Calm a Turbulent Sea." *Vikalpa,* Vol 28, No.1, Jan-Mar 2003, pp. 27–45.

Ray, Sougata (2004). "Environment-Strategy-Performance Linkages: A Study of Indian Firms during Economic Liberalization" Vikalpa • Vol 29 • No 2 • April - June 2004, pp 9-23

Rich, Robert and Tracy, Joseph (2003). "Modeling Uncertainty: Predictive Accuracy as a Proxy for Predictive Confidence." *Federal Reserve Bank of New York Staff Reports*, no. 161.

Sekharan, Uma (2003). *Research Methods for Business – A Skill Building Approach.* Fourth edition. John Wiley & Sons, Inc.

Walters, David (2003). "The Causal Texture of Organisational Environments Revisited"

Thompson, J D (1967). *Organizations in Action: Social Science Bases of Administrative Theory,* New York: McGraw-Hill.

## Planning and Forecasting

Arkes, M. R. (2001). "Overconfidence in judgmental forecasting," in J. S. Armstrong (Ed.) *Principles of Forecasting*. Norwell, MA: Kluwer Academic Publishers, pp. 495- 515.

Armstrong, J. Scott (1978). "Forecasting with Econometric Methods: Folklore versus Fact" *Journal of Business,* 51 (4), 1978, pp 549-564

Armstrong, J. Scott (1980). "The Seer-Sucker Theory: The Value of Experts in Forecasting." *Technology Review*, June/July, 1980, pp 16-24.

Armstrong J. Scott (1983). "Strategic Planning and Forecasting Fundamentals" in Kenneth Albert (ed.), *The Strategic Management Handbook*. New York: McGraw Hill, 1983, pp. 2-1 to 2-32.

Armstrong, J. Scott (1985). *Long-Range Forecasting: From Crystal Ball to Computer.* John Wiley, New York (2nd ed.).

Armstrong, J. Scott, Brodie Roderick and McIntyre, Shelby H. (1987). "Forecasting Methods for Marketing: Review of Empirical Research." *International Journal of Forecasting,* 3 (1987), 335-376, North Holland.

Armstrong, J. Scott (1989). Combining Forecasts: The End of the Beginning or the Beginning of the End? *International Journal of Forecasting* (1989), 5, pp. 585-588

Armstrong, J. Scott, and Collopy, Fred, (1992). "Error measures for generalizing about forecasting methods: Empirical comparisons," *International Journal of Forecasting,* 8, 69-80.

Armstrong, J. S. (2001). *Principles of Forecasting.* Boston: Kluwer Academic Publishers.

Armstrong, Scott J. (2001). "Standards and Practices for Forecasting". *Principles of Forecasting: A Handbook for Researchers and Practitioners.* Norwell, MA: Kluwer Academic Publishers.

Armstrong J. Scott (2002). "Assessing game theory, role playing, and unaided judgment", *International Journal of Forecasting* 18 (2002), pp 345–352.

Carbone, R. and J.S. Armstrong (1982). "Evaluation of extrapolative forecasting methods: Results of a survey of academicians and practitioners," *Journal of Forecasting*, 1, pp 215-217.

Clemen, Robert T. (1989). "Combining forecasts: A review and annotated bibliography," *International Journal of Forecasting*, 5, 559-583.

Collopy, F., J.S. Armstrong (1993). "Causal Forces: Structuring Knowledge for Time- series Extrapolation," *Journal of Forecasting*, 12 (1993), pp 103-115.

Collopy, F., J.S. Armstrong and M. Adya (1994). "Principles for, examining predictive validity: The case of information systems spending forecasts," *Information Systems Research*. 5, pp. 170-17

Coopersmith, L.W. (1983). "Forecasting time series which are inherently discontinuous," *Journal of Forecasting*, 2, pp 225-235.

Dalrymple, D.J. (1987). "Sales forecasting practices," *International Journal of Forecasting*, 3, pp 379-391.

Fildes, Robert (1985). Quantitative forecasting - the state of the art: Econometric models, *Journal of the Operational Research Society* 36, 549-580.

Fischoff, Baruch and D. MacGregor (1982). Subjective confidence in forecasts, *Journal of Forecasting* 1,155-172.

Gettys, C. F., Kelly, C., & Peterson, C. R. (1982). "The best-guess hypothesis in multistage inference", in Kahneman, D., Slovic, P. & Tversky, A. (eds.), *Judgment under uncertainty: heuristics and biases*. Cambridge, UK: Cambridge University Press.

Green Kesten C. and Armstrong, J.S. (2005). "Demand Forecasting: Evidence-based Methods." *Forthcoming Chapter in: Strategic Marketing Management: A Business Process Approach,* edited by Luiz Moutinho and Geoff Southern.

Harrison, P.J. and C.F. Stevens (1971). "A Bayasian approach to short-term forecasting," *Operational Research Quarterly*, 22, pp 341-362.

Kalwani, M. U. &. Silk, A. J. (1982). "On the reliability and predictive validity of purchase intention measures," *Marketing Science*, 1, 243-286.

Kumar V., Morwitz Vicki G. And Armstrong J. Scott (2000). "Sales Forecasts for Existing Consumer Products and Services: Do Purchase Intentions Contribute to Accuracy?" *International Journal of Forecasting*, 16, pp 383-397

Mentzer, J.T. and J.E. Cox (1984). "Familiarity, application and performance of sales forecasting techniques," *Journal of Forecasting*, 3, pp 27-36.

Mirchandani, R. (2007). *Development of a Demand Forecasting Model for Specialty Plant Nutrition Solutions applicable to India* [Doctoral Dissertation, NMIMS University].

Parthasarathy, N.S. (1994). *Demand Forecasting for Fertilizer Marketing.* Food & Agriculture Organization of the United Nations, Rome. FAO Corporate Document Repository. www.fao.org/docrep/003/t4240e/t4240e00.htm

Rowe, Gene and Wright, George (1999). "The Delphi technique as a forecasting tool: issues and analysis" International Journal of Forecasting 15 (1999), pp. 353–375

Schnaars, S.P. (1986). "Situational factors affecting forecast accuracy," *Journal of Marketing Research*, 21, pp 290-297.

Tsay, R. (1988). "Outliers, level shifts, and variance changes in time series," *Journal of Forecasting*, 7, pp 1-20.

Witt, S. and C. Witt, (1992). *Modeling and Forecasting Demand in Tourism.* Academic Press, London.

Yokum Thomas J. and Armstrong J.S. (1995). "Beyond Accuracy: Comparison of Criteria Used to Select Forecasting Methods" *International Journal of Forecasting,* 11, pp 591-597.

Yokum, J. Thomas, Collopy, F., J.S. Armstrong (2004). "Decomposition by Causal Forces: A Procedure for Forecasting Complex Time Series," *International Journal of Forecasting,* 21, pp 25-36.

# ABOUT THE AUTHOR

### Dr Rahul Mirchandani, Chairman & Managing Director - Aries Agro Limited

Dr Rahul Mirchandani has three decades of experience as part of the Promoter Group at Aries Agro Limited. Listed among the *"Top 20 Global Indian Leaders of 2023"*, curated by World Brand Affairs and selected as the *"Most Influential Leaders of New India in 2021"* by CNN News 18, Rahul is ranked amongst the *"50 Most Influential Rural Marketing Professionals of India"* - 2019 by CMO Asia and one of the *"30 Most Innovative CEOs in India"* - 2014 by INC India.

He has pioneered several unique marketing processes and brand management tactics at Aries. These include launching India's first loyalty programme in agribusiness, India's first Agribusiness flash sale (earning a Limca Book of Records), insurance-based customer retention programmes for farmers, executing the first nationwide single day launch for a specialty fertilizer brand as well as several shock-and-awe brand building programmes to sustain and grow 130+ rural brands with zero mass media advertising. Many of these strategies are being documented and taught as case studies in customer relations management at leading Indian business schools. Forbes India has also chronicled many of these tactics in 'Aries Agro: Brand of Choice for Millions of Farmers' in December 2020.

Rahul holds a Doctorate in Management Studies from NMIMS University, Mumbai and is also a Chartered Financial Analyst (CFA) and holds an MBA from the University of Canberra, Australia. Rahul has delivered sessions on Innovation and Entrepreneurship at the Oxford University, UK and has lectured at over 50 B-Schools in India. A Past National Chairman of the Confederation of Indian Industry's Young Indians (CII-Yi), he is the architect of Yi's Farmers Net program and has served on the CII Agriculture, Innovation, International Policy and India@75 National Councils. He has been the Chairman for the Yi's Next Practices platform and has also chaired Yi's International Relations and Partnerships and is the Founder of the Commonwealth Alliance of Young Entrepreneurs- Asia

(CAYE-Asia). He is the recipient of the Bharat Ratna Rajiv Gandhi Yuva Shakti Award 2010 in recognition of outstanding achievements towards Youth Empowerment and Inclusive Growth.

Recognized as one of the world's foremost achievers in his field, Rahul is listed in the *Who's Who in the World 2013* and has been invited to Receptions at Buckingham Palace, the International Labor Organization, the Commonwealth Heads of Government meetings, High Commissioners Banquets, Association of MBAs Global Deans conference, amongst many others. Rahul was a member of the Global Jury for the UN-Habitat Youth Entrepreneurship and Innovation Awards 2016. Rahul is a Fellow of the 4Sight Class of Ananta Aspen's India Leadership Initiative and member of Aspen Global Leadership Network.

**To connect with the Author**
**LinkedIn**

**Login to use the**
**Forecasting Algorithm**
**EmpowerToEnslave.com**